Applying Theories for Information Systems Research

I0885355

This book sets out to provide postgraduate researchers with guidance on selecting and applying sociotechnical theories to the study of information systems, including how they can be combined to complement each other. Until now it has been difficult to source advice on the application of these theories, and there has been no single book that combines multiple theories as this does.

Examining the impacts of technological developments and seeking to understand how humans interact with computers and systems is a dynamic field but can often confuse researchers with the overwhelming number of social theories that are utilised to derive insights. Instead, the author in this book breaks down some of the most popular theories used to underpin information system research, such as activity theory (AT), actor-network theory (ANT), contingency theory (CT), diffusion of innovation (DOI) theory, structuration theory (ST), and the technology acceptance model (TAM). By doing so, this book serves to enable a simpler, faster selection of appropriate theories, and a more effective and productive application that leads to richer, more rigorous research outcomes.

Written for postgraduates, researchers, and academics in the fields of information technology and information systems, this book provides a valuable resource of sociotechnical research methodologies that will enable and enhance future studies.

Tiko Iyamu holds a PhD in information systems from the University of Cape Town, South Africa. Currently, Iyamu is Research Professor at the Faculty of Informatics and Design, Cape Peninsula University of Technology (CPUT), Cape Town, South Africa. Previously, he was with the Tshwane University of Technology, South Africa, and Namibia University of Science and Technology, Namibia. Iyamu served as a Professor Extraordinaire at the Department of Computer Science, University of the Western Cape, from 2006 to 2015. In 2014, he was a visiting professor at the Flensburg University of Applied Sciences, Germany. Professor Iyamu's areas of focus include enterprise architecture, health informatics, big data analytics, mobile computing, and the application of sociotechnical theories for information systems research. He has authored 5 books and over 140 peer-reviewed research articles in journals, conference proceedings, and book chapters.

Applying Theories for Information Systems Research

Tiko Iyamu

Routledge
Taylor & Francis Group
LONDON AND NEW YORK

First published 2022
by Routledge
2 Park Square, Milton Park, Abingdon, Oxon OX14 4RN

and by Routledge
605 Third Avenue, New York, NY 10158

Routledge is an imprint of the Taylor & Francis Group, an informa business

British Library Cataloguing-in-Publication Data
A catalogue record for this book is available from the British Library

Library of Congress Cataloging-in-Publication Data
A catalog record has been requested for this book

ISBN: 978-1-032-02600-8 (hbk)
ISBN: 978-1-032-02601-5 (pbk)
ISBN: 978-1-003-18411-9 (ebk)

DOI: 10.4324/9781003184119

Typeset in Times New Roman
by codeMantra

Contents

Figures

Tables

Preface

The development, implementation, and management of information systems and technologies (IS/IT) are increasingly complex and challenging, primarily because of factors such as human interactions, unprecedented human actions, emergence of IS/IT solutions, rapid changing of organisational needs for IS/IT solutions, including transformation and cultural affiliations, which aid the formation of networks within the computing environments. Thus, information systems (IS) studies require rigour and deeper and fresher views using different lenses if different results are to be achieved in addressing the problems with IS/IT in organisations.

The application of sociotechnical theories to underpin IS research continues to increase. This more often than not means that the theories are applied as lenses to guide analysis of data and interpretation of findings, which ultimately shape the final outcome of studies. Hence, selecting the appropriate theories to underpin a research is consequential. The book covers six, activity theory (AT), actor-network theory (ANT), contingency theory (CT), diffusion of innovation (DOI) theory, structuration theory (ST), and the technology acceptance model (TAM), which are the theories mostly applied in IS research. This book provides a guide, from three fundament fronts, on how to select a theory, apply the theories, and complementarily apply theories.

The rapidly changing nature of businesses and user requirements makes it more difficult to develop and implement systems within a budget and timeframe. Many organisations focus on the utilisation of IS/IT solutions, and others are more interested in producing and reproducing technology over time and space through innovations. What is even more important but challenging is how the technologies are diffused and used or not used. DOI helps to understand the practice and philosophical assumptions of the symmetric between technology innovation and the social context, while TAM focuses on understanding the factors that impact on the acceptance or rejection of information technologies in organisations. AT is more concerned with how actors come together to execute tasks. ANT concentrates on power relationships among actors within heterogeneous networks, while ST focuses on the relationships and interaction between agents.

Sociotechnical theories have been employed to underpin many information systems studies over the years. Even though these theories have been used in many IS studies over the years, their application remains difficult. Also, the understanding that an interpretive tradition and approach implies a subjective epistemology and the ontological belief that reality is socially constructed makes the application of the theories very important but challenging. The ease of selecting and applying the theories are the main motivations and contributions of this book.

Acknowledgements

I am thankful to the Department of Information Technology, Cape Peninsula University of Technology. Graciously, I extend my humble appreciation to those with the Research Forum at the IT department and the postgraduate students that I have been fortunate to work with over the years. Tony Ekata, I thank you immensely for painstakingly proofreading the manuscript.

I feel a deep sense of gratitude to my siblings, particularly Ikpomwenosa (IK) and Austin, for their constant love and interest in what I enjoyably do. I am deeply indebted to my family for their consistent and constant love and care; Ruvimbo, thank you for your patience and support while I burn the midnight candle.

Gratefully, I offer warmth and profound appreciation to my older cousin, IK Ediae, and my uncle, Thomas Odemwingie, for their continued guidance and support. Very importantly, I am privileged to have Fred Eboka, Tony Uwagbafor, and Osagie Omoregbee as friends. Thank you for always being there for me.

I dedicate this book to my son, Osa, mother and late father, Mr and Mrs Iyamu.

1 Introduction

Table of contents

Introduction

I would like to begin by clarifying very important terms used in the chapters of this book: information systems and technology (IS/IT) and information technology (IT) solutions mean the same thing in the context of this book; also, environment and organisation mean the same thing. Thus, the terms are interchangeably used in this book.

The introductory chapter is divided into four sections: the introductory section, followed by a synopsis of the problem the book tries to address, which is presented in the second section. The third and fourth sections provide brief introductions of the information systems (IS) discipline and IS research. These sections highlight the significance of qualitative data and the essentiality of the data source, as detailed in Chapters 2 and 3. In the fifth section, six sociotechnical theories that are the focal point of this book are introduced. The theories are extensively and comprehensively covered in the last nine chapters of the book.

The goal and synopsis of the problem

Many challenges are implicitly associated with over-reliance on an interpretivist approach for analysis and interpretation of qualitative data in IS studies. Apparently, this is the main reason many IS researchers struggle to

DOI: 10.4324/9781003184119-1

justify the rigour in their studies. It gets worse because the data sets are often small, compared to quantitative studies. The emergence of sociotechnical theories has been significant in improving rigour and justification in IS research. Within the last two decades, six of the theories have been applied more than 5 million times in IS studies. In that period, last two decades, two things have been achieved and are well documented in literature: (1) the theories are well described and explained, including critique and support; and (2) the theories have been well embraced and are increasingly applied in IS studies.

Despite these efforts, a fundamental gap exists, which is the way the theories, can be applied in IS studies. The challenges emerge from two different perspectives; namely, why and how the theory is applied in an IS study. Some of the challenges emanate from the lack of know-how. This discourages postgraduate students from selecting a theory, even though it would have been most appropriate for their studies. This is primarily because the aspect that is found to be challenging and pushes some researchers, including postgraduates, away from the complexity of using the theories. This book provides a guide for each of the theories that has been most selected and focused on by IS scholars. The theories are activity theory (AT), actor-network theory (ANT), contingency theory (CT), the diffusion of innovation (DOI) theory, structuration theory (ST), and the technology acceptance model (TAM).

For each of the theories, this book proposes a model that holistically guides data analysis from an interpretivist perspective. Through the models, better understanding can be gained about the components of each theory and how they relate to the activities in the phenomena being studied. This includes the process of data collection in qualitative research and how its challenges can possibly be avoided. The chapters in the book reveal and discuss how the theories are used as lenses to guide data analysis and interpretation from a qualitative research perspective in IS. Another contribution of the book is in its plausible explanation: the strength of the theories lies in how to connect human actions with the different aspects of technology solutions, including the interaction and relationships that happen in the course of development, implementation, management, and use of information technology (IT) solutions. Subsequently, this draws the attention of both IS practitioners and academics, particularly aspiring researchers and postgraduate students, to the usefulness of applying sociotechnical theories in IS research.

Information systems

Neither this chapter nor the book intends to redefine IS as a discipline. I only describe it based on my experience that IS is an academic discipline which constitutes two fundamental embodiments – information and systems – which are of pivotal importance to society and scholarly domains.

Information is from the Latin word of *information-*. In the Oxford English dictionary, it is defined as knowledge acquired through experience or study, knowledge of specific and timely events or situation. Simplistically, a system consists of organised components that have relationships which go through integrated processes of different elements. The field of IS is a sociotechnical organisational system that is designed to gather, store, process, and distribute information (Ritchie, 1991). IS are made up of four main components: task, people, structure (or roles), and technology. None of the components operates in a vacuum; they influence one another to produce results or enhance usefulness. How, when, and why the components influence each other is determined through research.

In practice, people often change the IT solutions that they make use of by consciously or unconsciously modifying the properties through their interaction with the solutions. This happens even when a technology seems to be stable and functions appropriately. Thus, the stability of IT solutions and their application are only provisional. Consequently, because of the rapid evolution of IT solutions, which stems from the fact that different elements are continuously developed, existing functions fail and repairs are needed, new materials are invented, new standards are set, and users modify the artefact and/or its content for new and different uses. Thus, an understanding of how and why things evolve is key to findings in IS research. As has been seen or experienced for many years, there is increasing interest and the use of sociotechnical theories as lenses to examine and gain a better understanding of how and why things happen in the way that they do, in the development and implementation of IT solutions.

IS research is multifaceted and interdisciplinary, and is concerned with the study of the impacts and effects of information and systems on the behaviour of actors, agents, and organisations (Galliers, Markus & Newell, 2006). Hevner et al. (2004) explain how research in IS is used to create new solutions for both human and organisational capabilities. The solutions are constructs in the forms of models, frameworks, and methods (March & Smith, 1995). In addition, research is conducted for the purposes of innovation and contribution to knowledge. Essentially, research is vested in data and in the analysis of the data.

The analysis of data is aimed to answer research questions and to provide insights into solving problems. A research problem is not necessarily something that is broken. It could be enhancement, demonstration of knowledgeability, revelation of the unforeseen, or empirical confirmation of situations or circumstances. Data analysis is therefore a systematic approach which employs logical reasoning or techniques to describe, evaluate, and interpret data to find meanings that alter the ontological position before the research was conducted, and conclusively contribute to knowledge. Also, analysis is aimed to evaluate and examine sets of data, and to gain insights and an understanding on why and how systems' components interact, interconnect, interrelate, and influence each other in the process. Substantially, this is often a complex exercise; therefore, it requires symmetric, asymmetric, and isometric logical interventions in achieving the aim and objectives of the research.

IS research

The reliance on IS makes it critical to understand the various ways in which IT solutions can be used to support and enable organisations' operations. However, the selection, development, and implementation of IT solutions are not as simple as they are sometimes proclaimed, hence the existence of continuous challenges and the subsequent need for research. Owing to their complexities, from both technical and non-technical perspectives, organisations and researchers employ a more innovative approach, such as the use of theories, to underpin IS studies. Also, the use of theories add rigour, which improves reliability, credibility and legitimacy of the study. According to Rose and Scheepers (2001), theories contribute significantly to the development of the IS discipline because they help to gain a better understanding of human interactions with technologies in order to have a meaningful impact in an environment.

The interactions and relationships that occur among actors in the activities of IS are often unpredictable and sometimes unstructured, which makes it subjective in nature. However, subjectivism in a qualitative study does not necessarily make the examining and understanding of IT solutions easy. Qualitative studies in IS primarily focus on actors' relationships and interactions, from the perspective of subjectivism. Callon (1986) defines an actor as being both a human and a non-human entity. According to Myers and Avison (2002: 70), 'qualitative research methods were developed in the social sciences to enable researchers to study social and cultural phenomena'. The primary purpose of qualitative research is to understand a phenomenon as it is seen by respondents within contexts and over a period of time. Purposefully, this is achieved by studying the respondents' views in the context of their natural settings. The outcomes from qualitative research are subject to the meanings which people give to them in real-life situations and contexts (Yin, 2010). However, the meaning which individuals and groups give or associate with events has never been easy for researchers and practitioners, without analysing or interpreting them.

Data analysis is critical in empirical studies, in that it entails the unpacking of data into perspectives (Bryman, 2012). Analysis is a process of making sense of the data that was gathered within context and the relevance of the phenomena being studied. The relevance of data is determined by the research questions and objectives. Increasingly, sociotechnical theories are employed to guide data analysis in an interpretivist study. There are two main components of theory: relevance and context. Relevance is symbolically representative of real-life issues or challenges. Context is deterministic of scope and boundaries. The components are viewed from different elements such as: (1) some theories are more complicated in nature than others;

(2) one or a combination of theories can be applied in a research; (3) theories add rigour or quality to research; and (4) theories serve important purposes in that they shape the outcome of research.

The use of theories in IS tradition began around the mid-1990s. Since then, interest from various quarters of IS has grown significantly (Mueller & Urbach, 2013). The theories are employed to facilitate knowledge in the process of extracting findings. In the process, it has helped produce high-quality scholarly contributions (Tsang & Ellsaesser, 2011). Also, it helps to advance an understanding through explicit explanation of the phenomenon being studied. Another issue of pivotal importance is that the use of theories increasingly adds legitimacy to findings and conclusions reached in IS studies.

Sociotechnical theories are vital in IS research in that they guide and give meaning to humans' experiences, feelings, thoughts, and knowledgeability. They guide the explanation of what, when, why, and how factors influence actors or agents in the phenomenon being studied. Consequently, they add rigour, which ultimately helps to shape the outcome of the study. On the one hand, they virtually extend the boundaries of the known and unknown of existing knowledge. On the other hand, they bring about challenges from the angle that a good theory is not necessarily a useful theory in a study. The rhetorical question that immediately follows is, what makes a theory useful in a study?

IS researchers formulate and use various criteria to evaluate the usefulness of theories in the discipline. The usefulness of sociotechnical theory in IS research depends on the nature of the phenomenon being studied, which is based on the research problem or the objectives. There are four fundamental elements: (1) the context of the study; the theory is purposely applied to give meaning to circumstantial incidents and events; (2) resonance with the objectives, to guide the data analysis towards answering the research questions; (3) ability to assist in the explanation of the findings that corroborate with the actions and interaction of the actors or agents within the context of the research objectives; and (4) ability to stimulate the development of a theoretical concept (framework or model) in the form of new knowledge.

There is emphasis on the use of theories from two main angles: (1) as a manner of speaking, which implicitly means research language; and (2) as a way of thinking, which implies unprecedented or a beyond-the-norm type of approach or end. Both elements are amplified in the use of theories as lenses. First, as a manner of speaking, it helps to follow the traces in relationship and interaction that happen between actors. Second, as a way of thinking, it is an integral part of research, primarily because it is used to gain a deeper and better understanding of what happens, how things happen, and why things happen in the way that they do. It is based on this premise that it helps to reach findings, explain the findings, give meaning to findings, and draw conclusions.

There are many theories and they exist in the areas of psychology, sociology, organisation, economy, and system (technical). This book focuses on sociology and technical (sociotechnical) theories that have increasingly been applied in IS research over the last two decades. Specifically, this book concentrates on the most popular sociotechnical theories in IS research, which are AT, ANT, CT, DOI theory, TAM, and ST. As presented in Chapter 10, AT has been applied over 3 million times, followed by DOI, ANT, and ST. The least of the top six are CT and TAM, which have been applied in IS studies over 60,000 and 70,000 times, respectively.

A lens is an aspect of a theory that converges on a particular perspective through which analysis can be conducted to examine and explain a phenomenon. It thus, specifically, focuses on links, traces, patterns, relationships, and interactions between actors in the development, implementation or management of IT solutions. Based on the objectives of the study, a lens is used to view a given situation to explain the world view explicitly or implicitly. It is thus used to explain the constituent elements and the interrelations in findings.

Sociotechnical theories in IS

The sociotechnical theories emphasise on social contexts from different angles and offer various types of meanings and explanations. The theories bring fresh perspectives into IS research. The theories are applied to facilitate and examine the interactions and relationships that exist between actors or agents, and between people and technology. This includes the networks which the actors form under conscious and unconscious circumstances, while developing, implementing, using, or managing IT solutions for organisations' purposes. Thus, the use of the interpretivist approach by itself may not be exhaustive enough in the analysis of qualitative data that can be so complex in IS research. The multifaceted nature of qualitative data rationalises, in addition, the increasing use of sociotechnical theories in the field of IS.

The remainder of this book is structured into 11 chapters. Chapter 2 covers the types and sources of qualitative data. The chapter that follows provides a detailed discussion on collecting qualitative data, focusing on the use of semi-structured technique. Chapters 4–9 sequentially cover applying AT; ANT; CT; DOI; ST; and TAM, respectively. Chapter 10 provides a guide on how to select a theory for an IS research. In Chapters 11 and 12, ANT and ST, and AT and ANT are combined, respectively.

Activity theory

AT was developed in the twentieth century for psychological purposes, which has since evolved and is being applied in IS studies (Yamagata-Lynch, 2010). The theory focuses on human activities within a socio-environment. It holds that the elements of activities are not fixed and can change as circumstances change over time or within space (Kaptelinin & Nardi, 2006). Engeström et al. (1999) explain how human activity is composed of three levels: activity, action, and operation. In the context, an activity comprises of actions, and each action endorses operations. Through this composition, AT is used to investigate human associated engagement or event that is understood as activity in a specific social setting. Evidently, activities evolve in a non-linear way. This makes AT a useful theory for explanation and in gaining a deeper understanding of the logic behind human actions in carrying out an activity.

Also, the theory helps to trace and examine how each activity is performed through group or individual actions or chains of actions that are related to one another towards achieving a specific goal or objective (Nardi, 1996). In AT, actions are conscious, and consciousness is a basic principle of AT in seeking to understand actions and outcomes (Kaptelinin & Nardi, 2006). Consciousness is described by Kaptelinin and Nardi (2006: 9) as 'the enactment of our capacity for attention, intention, memory, learning, reasoning, speech, reflection and imagination'. AT consists of six components: (1) tools, (2) subject, (3) object, (4) rules, (5) community, and (6) division of labour, a subject and an object mediated by a tool (Engeström et al., 1999). The components are comprehensively discussed in Chapter 4. In IS, AT is often used to account for history and is particularly useful as a lens in qualitative studies where methodologies such as ethnography and case study are followed. It guides analysis to find patterns of actions in interactions and relationships, describing phenomena and presenting phenomena through a built-in language and rhetoric.

Actor-network theory

ANT is basically a social theory that is embedded within science and technologies. The core elements of ANT are the actor, network, and translation (Callon, 1986), others are blackbox, irreversibility, and immutable mobile. In ANT, actors are both human and non-human (Callon, 1991). Neither actor nor network can exist separately or independently of each other; the relationship happens in a recursive process through which the network matures or becomes stable (Heeks & Stanforth, 2015). Translation is the process of movement from one stage to another and it entails negotiations between the actors involved. There are four stages in translation referred to as moments of translation (Callon, 1986). Many IS studies have benefited from the use of moments of translation as a lens for analysis and interpretation of qualitative data. ANT provides a lens to view how technology and social factors shape each other, which are often applied in the development, implementation or management of IT solutions.

ANT is concerned with the existence of social and technical factors, and the creation and maintenance of networks of human and non-human actors, which, in the case of IS, includes people, organisation, software, and hardware (Walsham, 1997). In ANT, the constituents of the network are constantly shifting, including the relationships between actors (Callon, 1986). This is vital in investigating the complexity that often exists in the relationship and interaction between actors in networks (Heeks & Stanforth, 2015). In IS research, ANT is vital in four main ways – investigating how: (1) networks come into being (Callon, 1986); (2) associations or relationships exist; (3) actors are enrolled into a network (Doolin & Lowe, 2002); and (4) a network or networks become stable. It is thus useful in investigating social and technical factors that are involved in the development, implementation, and maintenance of IT solutions in an environment.

Contingency theory

CT originates from organisational theory (Galbraith, 1973). However, it also derives from systems concepts, based on the notion that organisations comprise of a multitude of subsystems whose interrelationships have to be recognised (Reinking, 2012: 726). It describes the relation between the contextual factors, such as technology, the environment, and organisational structures, including the variables (Opitz, Krüp & Kolbe, 2014). Lin and Wang (2012) explain how relevant factors influence IT solutions and are perceived fit by denoting internal and external contingencies. The theory provides four variables – environment, strategy, structure, and performance – that are increasingly used in IS research (Blanton, Watson & Moody, 1992).

The theory is based on two fundamental principles. First, there is no best way to organise or manage an organisation rather influenced by uncharacteristic internal and external factors. Thus, each situation should be seen as unique and requires distinctiveness in order to achieve optimal functionality (McGrandle, 2017). This makes the theory most appropriate in examining the various best ways of constructing effective and efficient IT solutions. Second, any specific method that an organisation employs to organise or manage its activities is not equally effective (Van de Ven, Ganco & Hinings, 2013). This means that the optimal course of action is contingent on internal and external factors.

Diffusion of innovation

DOI theory seeks to explain how and why including the rate at which a new idea or technology is dispersed in an environment. Vitally, this helps to reveal factors of influence in assessing the likelihood of success or failure in the development or implementation of IT solutions. Rogers (2003) argues that diffusion is the process of disseminating an innovation within a social system. In DOI, four components – innovation, communication channel, time, and social system – influence the spread of a new idea in an environment. According to Nemutanzhla and Iyamu (2011: 244), the 'DOI theory communicates technological innovation through specific networks, over time, among members of society'. Individuals and groups possess different degrees of willingness to adopt or not to adopt innovations, which are influenced by factors including time (Rogers, 2003). The theory is thus concerned with the manner in which a new technological idea, artefact, or technique, or a new use of an old one, migrates from one point (creation) to another (use).

Diffusion refers to the process by which innovations are spread among the members of a social system over time (in organisations), whereas adoption is a decision in implementing innovations based on knowledge and persuasion of individuals within a given system (e.g., organisations or enterprises) (see Rogers, 1995). The theory can be used at both macro and micro levels to assess or evaluate the adoption of innovation. Diffusion is at the macro level. It concerns the dissemination of an innovation, while the micro level is the adoption stage where an individual or group decides to accept or reject the new technology solution. The DOI process consists of five components: (1) knowledge, the first step in the process where awareness is required; (2) persuasion, a point at which a prospective adopter shows interest in the new technology; (3) decision, based on certain factors an adopter decides; (4) implementation, operationalisation of the innovation; and (5) confirmation, finalisation of the entire process.

Structuration theory

ST is a social theory that focuses on the reproduction of social systems through interaction between agents and structures (Giddens, 1984). In ST, agents are both technical and non-technical artefacts, and structures are rules and resources (Iyamu & Roode, 2012). The theory holds that human actions are enabled and constrained by structures but emphasises that these structures are the result of previous actions (Orlikowski, 1992). In structuration, structure is only manifested in the structural properties of social systems and consists of the rules and resources that human agents use in their everyday interaction. These rules and resources do not exist independently of human action, nor are they material entities. Giddens (1984) describes them as 'traces in the mind' and argues that they exist only through the action of human beings. According to Loureiro-Koechlin (2008), the agents in an organisation learn how to exercise their power as is expected in the social system they are in.

Agents create structures, which in turn enable and constrain their actions and interactions, referred to as duality of structure, the aspect of the theory that is mostly used as a lens in IS studies (Jones & Karsten, 2008). The duality of structure consists of three dimensions, namely, structure, modality, and interaction (Giddens, 1984). The interaction between agents and structures are linked through modality, interpretive scheme, facility, and norm. Signification is established by human agents based on the interpretative schemes through which the meaning of communication is affirmed. Human agents dominate through the use of facilities that are then used to exercise power. The theory is applied as a lens in many IS studies, primarily to guide analysis, particularly from the perspective of qualitative methods (Jones & Karsten, 2008).

Technology acceptance model

TAM is a theory that models how users get to a point where they accept or reject a technology solution. TAM is intended to provide an explanation on the determinants which influence the acceptance and rejection of systems within an organisation (Davis et al., 1989). TAM assumes that beliefs about usefulness and ease of use are always the primary determinants of the adoption of information technologies in organisations. The theory suggests that these two determinants serve as the basis for attitudes towards using a particular system which, in turn, determines the intention to use, and generates the actual usage behaviour (Venkatesh et al., 2003). Thus, the model is suitable for examining the factors which cause people to accept or reject IT solutions in organisations.

TAM provides the basis for examining the impact of external variables on internal beliefs, attitudes, and intentions (Davis, 1989). The use of TAM in IS studies helps to reveal valuable insights on why and how IT solutions are accepted or rejected by users in an environment. In TAM, there are two main determinants of technology acceptance: perceived usefulness and perceived ease of use (Wallace & Sheetz, 2014). Perceived usefulness and ease of use are influenced by external factors based on the characteristics of the technology.

Qualitative studies increasingly rely on sociotechnical theories to guide analysis and illuminate findings. In some studies, theories are used to guide the design of research questions, interpret qualitative data, and propose explanations for the insights of the influencing factor. The theories are used to provide meanings beyond an interpretivist approach. In addition, they help to provide an understanding of complex things, data, and circumstances such as the relationship between social system and technology solutions by underpinning: (1) how technologies are developed and implemented; (2) why social systems work in the way that they do; and (3) why humans interact in certain ways.

As briefly presented above, there are different lenses that can be employed to examine these circumstances in the various phenomena. This affirms that there is no one way to examine and understand how and why certain problems exist or why some IT solutions behave in the way that they do. Theories are used to underpin studies, to provide insights, and, therefore, to add rigour. Sociotechnical theories are vital in IS research, in that they guide and give meaning to human experiences, feelings, thoughts, and knowledgeability. Thus, they extend the boundaries of the known and unknown of existing knowledge.

The selection and application of theories are influenced by various factors primarily because they constitute integrated sets of attributes from different angles: (1) some theories are extreme in nature; (2) one or a combination of theories can be applied in a research; (3) theories add to the quality of data and the research rigour; and (4) theories serve important purposes in that they shape the outcome of research. Thus, two or more of the theories could be complementarily used in a study. Such a case requires different levels of analysis, with each theory having its focus. Iyamu and Roode (2012) employed both actor-network theory and ST to examine the factors which influence IT strategy in a financial institution. The aim was not to compare and contrast the two theories, but to use them in a complementary fashion. Chapter 10 addresses the need and how theories can be complementarily used in a study.

Conclusion

This is the introductory chapter. It briefly explains how theories are used to underpin IS studies to explain humans' and technologies' traits, and the interconnectedness between actors in phenomena being studied. The chapter introduces why sociotechnical theories should be used to guide explanation on what, when, why, and how factors influence actors or agents in the phenomenon being studied. Consequently, this rigorously helps to shape the outcome of the study.

The analysis aspect focuses on people's subjectivity as they continue to make use of technology for different reasons. The introductory content indicates that theories can reveal and facilitate the stability of an IT environment and that continuous change of technologies is subject to human actions. Hence, examining human actions and their manifestation is critical, using different sociotechnical theories. As discussed in the remainder of the book, theoretical lenses are intended to assist researchers to gain a deeper understanding of the phenomena that they study.

References

Blanton, J. E., Watson, H. J., & Moody, J. (1992). Toward a better understanding of information technology organization: A comparative case study. *MIS Quarterly, 16*(4), 531–555.

Bryman, A. (2012). *Social research methods* (4th ed.). New York: Oxford University Press.

Callon, M. (1986). Some elements of the sociology of translation: Domestication of the scallops and the fisherman of St Brieuc Bay. In: J. Law (ed.), *A new sociology of knowledge, power, action and belief*, pp. 196–233. London: Routledge.

Callon, M. (1991). Techno-economic networks and irreversibility. In: J. Law (ed.), *A sociology of monsters. Essays on power, technology and domination*, pp. 132–164. London: Routledge.

Davis, F.D. (1989). Perceived usefulness, perceived ease of use, and user acceptance of information technology. *MIS Quarterly*, *13*(3), 319–340.

Davis, F.D., Bagozzi, R.P., & Warshaw, P.R. (1989). User acceptance of computer technology: A Comparison of Two Theoretical Models. *Management Science*, *35*(8), 982–1002.

Doolin, B., & Lowe, A. (2002). To reveal is to critique: actor–network theory and critical information systems research. *Journal of Information Technology, 17*(2), 69–78.

Engeström, Y., Miettinen, R., & Punamäki, R.-L. (Eds.). (1999). *Perspectives on activity theory.* Cambridge: Cambridge University Press.

Galbraith, J. (1973). *Designing complex organizations.* Boston, MA: Addison-Wesley Longman Publishing Co., Inc.

Galliers, R. D., Markus, M. L., & Newell, S. (2006). *Exploring information systems research approaches: Readings and reflections.* New York: Routledge.

Giddens, A. (1984). *The constitution of society: Outline of the theory of structuration.* Cambridge: Polity Press.

Heeks, R., & Stanforth, C. (2015). Technological change in developing countries: Opening the black box of process using actor–network theory. *Development Studies Research*, *2*(1), 33–50.

Hevner, A. R., March, S. T., Park, J., & Ram, S. (2004). Design science in information systems research. *MIS Quarterly*, *28*(1), 75–105.

Iyamu, T., & Roode, D. (2012). The use of structuration theory and actor network theory for analysis: Case study of a financial institution in South Africa. In: A. Tatnall (ed.), *Social influences on information and communication technology innovations*, pp. 1–19. IGI Global, Hershey, Pennsylvania, USA.

Jones, M. R., & Karsten, H. (2008). Giddens's structuration theory and information systems research. *MIS Quarterly*, *32*(1), 127–157.

Kaptelinin, V., & Nardi, B. (2006). *Acting with technology: Activity theory and interaction design.* USA: Massachusetts Institute of Technology (MIT) Press.

Lin, W. S., & Wang, C. H. (2012). Antecedences to continued intentions of adopting e-learning system in blended learning instruction: A contingency framework based on models of information system success and task-technology fit. *Computers & Education*, *58*(1), 88–99.

Loureiro-Koechlin, C. (2008). A theoretical framework for a structuration model of social issues in software development in information systems. *Systems Research and Behavioural Science: The Official Journal of the International Federation for Systems Research*, *25*(1), 99–109.

March, S. T., & Smith, G. F. (1995). Design and natural science research on information technology. *Decision support systems*, *15*(4), 251–266.

McGrandle, J. (2017). Understanding diversity management in the public sector: A case for contingency theory. *International Journal of Public Administration*, *40*(6), 526–537.

Mueller, B., & Urbach, N. (2013). The why, what, and how of theories in IS research. Thirty Fourth International Conference on Information Systems, Milan.

Myers, M., & Avison, D. (2002). *An introduction to qualitative research in information systems: A reader.* Southampton: Sage Publications.

Nardi, B. (Ed.). (1996). *Context and consciousness: Activity theory and human-computer interaction.* Cambridge: Massachusetts Institute of Technology (MIT) Press.

Nemutanzhela, P., & Iyamu, T. (2011). A framework for enhancing the information systems innovation: Using competitive intelligence. *The Electronic Journal of Information Systems, 14*(2), 242–253.

Opitz, N., Krüp, H., & Kolbe, L. M. (2014). How to govern your Green IT?-Validating a contingency theory based governance model. In *PACIS* (p. 333).

Orlikowski, W. J. (1992). The duality of technology: Rethinking the concept of technology in organizations. *Organization Science, 3*(3), 398–427.

Reinking, J. (2012). Contingency theory in information systems research. In: Y. Dwivedi, M. Wade, & S. Schneberger (eds.), *Information systems theory*, pp. 247–263. New York: Springer.

Ritchie, L. D. (1991). Family communication patterns: An epistemic analysis and conceptual reinterpretation. *Communication Research, 18*(4), 548–565.

Rogers, E. M. (1995). *Diffusion of innovations* (4th ed.). New York: Free Press.

Rogers, E. M. (2003). *Diffusion of innovations* (5th ed.). New York: Free Press.

Rose, J., & Scheepers, R. (2001). Structuration theory and information system development-frameworks for practise. *In the proceedings of the 9th European Conference on Information Systems*, (pp. 217–231). Bled.

Tsang, E. W., & Ellsaesser, F. (2011). How contrastive explanation facilitates theory building. *Academy of Management Review, 36*(2), 404–419.

Van de Ven, A. H., Ganco, M., & Hinings, C. R. (2013). Returning to the frontier of contingency theory of organizational and institutional designs. *Academy of Management Annals, 7*(1), 393–440.

Venkatesh, V., Morris, M.G., Davis, G.B., & Davis, F.D. (2003). User acceptance of information technology: Toward a unified view. *MIS Quarterly, 27*(3), 425–478.

Wallace, L. G., & Sheetz, S. D. (2014). The adoption of software measures: A technology acceptance model (TAM) perspective. *Information & Management, 51*(2), 249–259.

Walsham, G. (1997). Actor network theory and IS research: Current status and future prospects. In: A. S. Lee, J. Liebenau, & J. I. DeGross (eds.), *Information systems and qualitative research,* pp. 466–480. London: Chapman & Hall.

Yamagata-Lynch, L. C. (2010). Understanding cultural historical activity theory. In: L.C. Yamagata-Lynch (ed.), *Activity systems analysis methods*, pp. 13–26. Boston, MA: Springer.

Yin, R. (2010). *Qualitative research from start to finish.* New York: Guilford Press.

2 Types and sources of qualitative data

Table of contents

Introduction

Research is an investigative practice to reveal the unknown, or to empirically substantiate the known, using a vehicle referred to as research methodology. Research is thus a systematic process that involves collecting, analysing, and interpreting data in order to increase our understanding of a specific phenomenon and thereby contribute to the body of knowledge (Leedy & Ormrod, 2010). According to Mouton (2001: 35), research methodology focuses on 'the knowledge of how or the "know-how" to do things or the total set of "means" that scientists employ in order to reach their goal of valid knowledge'. Research methodology consists of methods; popular among them in information systems (IS) studies are qualitative and quantitative methods described as the overall approach in a research (Myers, 1997). In addition, the two methods can be combined and referred to as mixed method. This chapter focuses on the qualitative method.

The qualitative research method is a field of enquiry which can be explored through different paradigms, such as interpretivism. According to Kelliher (2005), qualitative research is a constructivist methodology which follows a post-modernist and naturalistic approach and is often associated with the interpretivist approach in the analysis of data. The interpretivist approach guides the meanings that are associated with things to gain a deeper understanding of how human agents or actors in the social context interpret and execute their cognitive understanding towards their own reality. Also, the interpretivist approach is used to gain an understanding

DOI: 10.4324/9781003184119-2

of the inter-subjective meanings embedded in social life, and to explain why people act in the way they do. Thus, the interpretivist approach by itself is likely to fall short in an extremely complex case, thereby challenging its use for data analysis. It is in such situations that the use of theories to underpin the study is critical. This is to increase the depth of empirical works and strengthen the study in a more technically advanced adaptation and to gain a better understanding of the relationship and interaction between humans and technologies.

In order to comprehend human actions, it is essential to understand their motives. This means that there must be some common ground upon which the researcher and the research object can agree on meaning. In qualitative studies, understanding and examining human actors' or agents' actions is guided by philosophical assumptions. Thus, the ontological and epistemological bases implied by the underpinning theories, such as the actor-network theory, the structuration theory, the diffusion of innovation theory, and the technology acceptance model are used. The theories are used in two main fundamental ways: as a way of thinking and speaking about the phenomena being studied and as a 'lens' through which data is viewed and interpreted. Thus, they are used to make sense and provide explanations to problems. According to Kelly (2009), a theoretical underpinning in qualitative studies is often underplayed but it is very important, as it improves the quality of research from the perspective of design, interpretation, and analysis of data.

This chapter covers the types and sources of data within the qualitative methods in IS research. This includes strategy and the techniques that are applied in empirical studies. The chapter is organised into two main sections: philosophical assumption and methodology. The first section is concerned with philosophical assumptions and a philosophical stance, which are interpretivism and positivism. The second section discusses research methodology, which includes research design, data collection, and data analysis. At the end, the chapter is summarised.

Philosophical assumption

This section introduces philosophical assumptions in IS research but does not deal with the topics holistically. The topics have been comprehensively and consistently addressed over the years: See Hassan, Mingers, and Stahl (2018), Rowe (2018), Hirschheim and Klein (2012), and Dubé and Paré (2003). In studying phenomena, the philosophies are intertwined. IS research is based on philosophical assumptions, ontology, epistemology, or axiology.

Research is carried out to investigate an object or a subject or both, through philosophical assumptions which contain the vehicle (Methodology) that is employed to deliver the objectives. The study could be either applied or basic research. Applied Research is carried out with the intention to apply the results and findings to solve specific current or future problems. Basic Research is conducted with the intention to gain a better understanding and acquire more knowledge of the phenomena that occur, as well as to build theories based on the findings from the research. *Both types of research follow methodological steps of systematic enquiry, either through a quantitative or through a qualitative paradigm, or both, in order to arrive at solutions to problems.* Before the methods, the philosophical assumptions and stances are established.

Philosophical assumption is the first step in studying a phenomenon. Philosophy is an abstractive approach or postulation, while assumption is an idea that is moulded without evidence, or supposition, and often temporarily accepted as true. When the philosophy is made explicit, it reveals the assumption which informs or triggers research in IS. Common philosophical assumptions in IS research are ontology, epistemology, and axiology. Ontology refers to the things that exist and the properties that they do have, in the context of the social world. This forms the assumptions held about the nature and existence of social reality. Epistemology is concerned with the nature of knowledge and ways of knowing what can be known or learned about social reality. This enacts sense making that is associated with subjectivism or objectivism. This is distinct from ontology and epistemology.

The quantitative research paradigm is concerned with the quantification of constructs, thereby attaching numbers to the qualities of things (Babbie, 2007: 49). Researchers employ quantitative research to assess value and measure phenomena. Thus, quantitative research studies are not underpinned by social theories. In contrast to quantitative research, qualitative research is concerned with understanding and interpreting how humans create and maintain their social world rather than explaining and predicting behaviour, naturalistic observation rather than controlled measurement, and the subjective exploration of reality from the perspective of an insider as opposed to the outsider perspective. This type of research studies should be underpinned by social theories to examine and understand the relationship and interaction between humans in their association with systems and technologies.

In the second step, a decision is made about the philosophical stance, positivism, interpretivism, or critical realism. In relation to IS research, Orlikowski and Baroudi (1991) identify three philosophical perspectives. These are the positivist, interpretivist, and critical research perspectives. The stance (position) has an impact or implication on other aspects of the research, which include method, design, data collection, and analysis of the data. A research method can accordingly be positivist, interpretivist, or critical (Walsham, 1995). The next three paragraphs discuss these research approaches. The discussion is not detailed because it is the focus of neither this chapter nor the book. Also, these topics have been consistently discussed extensively in books and research articles in the last two decades. See positivism (Kock, Avison & Malaurent, 2017; Paré, 2004; Trauth & Jessup, 2000), interpretivism (Chen & Hirschheim, 2004; Goldkuhl, 2012; Walsham, 1995), and critical realism (Cecez-Kecmanovic, 2005; Mingers, Mutch & Willcocks, 2013; Smith, 2006).

Positivism in IS research is based upon the assumption that reality is objectively given and that it can be described by reference to measurable properties that are independent of the researcher (Myers, 1997). Also, the approach is deeply rooted in functionalism and too concerned with causal analysis at the expense of getting close to the phenomenon being studied (Galliers, 1991). In positivism, objectivism is a characteristic that resides in a phenomenon, which rejects the notion of how individuals see the social world as it 'really' is. Thus, the data being gathered must be accurate, meaning as objective as possible (Straub, Boudreau & Gefen, 2004). Therefore, an individual reasoning cannot be beyond a parameter, which means that the explanation of 'why' things happen is out of the equation. Theories are applied purposely for explanation and not for precise prediction. In applying sociotechnical theory, any individual can see the world from any or many angles as perfectly as it really is.

The interpretive research approach focuses on 'reality', but from a different perspective to that of the positivist approach. An interpretivist approach can help the researcher gain knowledge of reality through social constructs such as language, shared meanings, tools, documents, etc. (Walsham, 1993). Within the interpretive research paradigm, there are neither predefined dependent nor independent variables, which therefore require filtering of lenses. The focus is primarily on the complexity of human sense making of the subject as the situation emerges. The interpretivist approach refers to how knowledge about the world can be gained through human understanding and the meanings they associate with things. This means that human interest cannot be disassociated from the phenomenon being studied beyond objective evidence to include feelings and experiences. In Walsham's (1993) explanation, there is nothing correct or incorrect in the social world; rather, things are interesting or less interesting.

The critical research approach in IS research sees its main task as one of social critique, whereby the restrictive and alienating conditions of the status quo are shown and challenged (Klein & Myers, 1999). In critical research, the investigation is classified as emancipative if it aims to help eliminate the cause of unwarranted alienation and domination, and thereby enhance the opportunities for the realisation of human potential (Hirschheim & Klein, 1989). This is attributed to the fact that 'our knowledge of reality is a result of social conditioning and, thus, cannot be understood independently of the social actors involved in the knowledge derivation process' (Dobson, 2002: 1). Therefore, 'the nature of what exists cannot be unrelated to how it is studied' (Archer, 1995: 16). Thus, reality itself is a product of knowledge, which is an outcome from a study. Critical theorists assume that people can consciously act to change their social and economic conditions. They also assume that social reality is historically constituted and that it is produced and reproduced by people. Critical research assumes that people are constrained in their actions by various forms of cultural and political domination (Myers, 1997). The involvement of people's actions makes it essential to employ theory in determining why things happen in the way that they do.

The interpretivist approach is mostly associated with qualitative data, and often employs sociotechnical theories. Since this book is about applying sociotechnical theories in IS research, it is on that basis the focus is on an interpretivist approach to gain a deeper understanding of how and why social entities, such as information technology (IT) solutions, exist in the way that they do.

Research methodology

Research methodology consists of methods, techniques, and approaches within systematic steps, such as strategy and design, as shown in Figure 2.1. It is the vehicle that describes the entire scenario of the research.

Research strategy

The strategy defines the primary direction of the research, in that there are different types of methods such as qualitative and quantitative that can be adopted or followed. However, both qualitative and quantitative methods can be used complementarily as a mixed method. This section or chapter

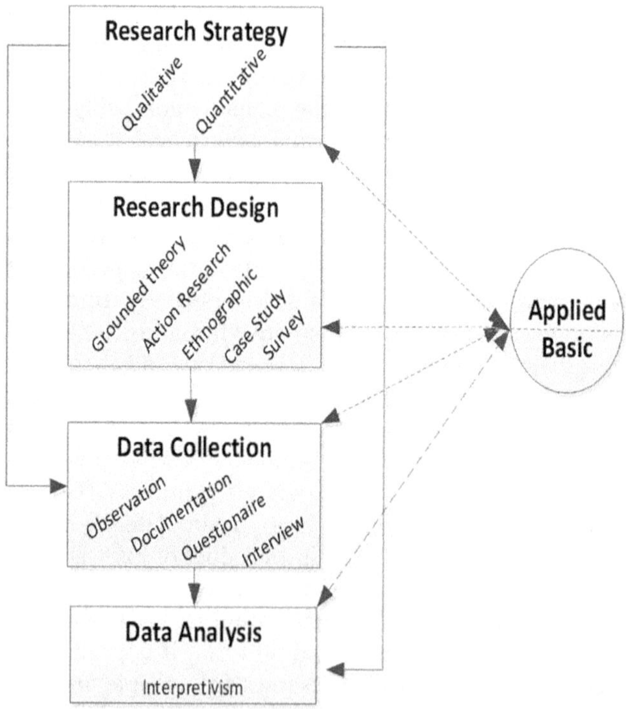

Figure 2.1 Research methodology.
Source: Author.

is not about qualitative or quantitative methods. However, the methods are covered in detail and in the context of the section the methods are distinguished at a high level. The primary difference between the qualitative and quantitative research methods is that qualitative research provides insights to an understanding of the problem in a natural setting, whereas quantitative research seeks to quantify the data and measure the value in statistical form.

The quantitative research method primarily focuses on the quantification of constructs, thereby associating numbers with the qualities of things (Babbie et al. 2007). The method is therefore considered and referred to as an enquiry procedure which seeks to confirm hypotheses by taking statistics and numerical figures into consideration. In Nenty's (2009) argument, quantitative research is conceptually made up of three sections: (1) verbal discourse; (2) numerical discourse; and (3) a section, again, based on verbal discourse.

Quantitative research measures variables on instruments so that numeric data can be analysed, using statistical procedures to seek explanations and predictions that can be used for generalisation (Leedy & Ormrod, 2010). Along the same line of argument, Creswell (2009) described quantitative research as a means for testing theories by examining the relationship among the variables. According to Gay and Airasian (2000: 25–26), the main aim of descriptive research in quantitative approaches is the exploration and clarification of phenomena where accurate information is lacking.

The qualitative method is typically used to investigate behaviour as it occurs naturally in a non-contrived situation, with no manipulation of conditions. According to De Vos et al. (2002), qualitative research aims at understanding the social world and the meaning that people attach to things that are around them. Most importantly, the method allows for close interaction with participants, such as interviewees, enabling the researcher to develop a deeper understanding of the situation that is being studied. One of the assumptions underlying qualitative research is that knowledge is socially constructed. Also, the qualitative research method allows a researcher to observe events and actions as they happen, without removing them from their context or natural settings (Babbie & Mouton, 2001). This means that the data consists of words in the form of rich verbal expressions and descriptions, rather than numbers and quantifications.

The qualitative method helps to enhance an understanding of the phenomena being studied through closeness with the subject. Also, it helps to describe the processes of knowledge sharing. This is attributed to the fact that the qualitative research method explores attitudes, behaviour, and experiences through methods like interviews, observation, or focus groups (Sekaran, 2003).

Even though the Mixed Methods is not covered in Figure 2.1, it is necessary to provide a brief about the method. It is the systematic integration of both qualitative and quantitative methods in a study. This implies the combination of qualitative and quantitative data during data collection, analysis, or discussion. Zachariadis, Scott, and Barrett (2013) explain how the mixed method enables addressing the issues of validity and by analysing combined data from both qualitative and quantitative methods, it richly contributes to a distinctiveness theory.

The methods are essentially useful in gaining an understanding of contradictions between quantitative results and qualitative findings. It is applied to improve an evaluation by ensuring that the limitations of one type of method are balanced by the strengths of the other. Thus, it provides rich insights and an opportunity to develop novel theoretical perspectives. Venkatesh, Brown, and Sullivan (2016) suggest that by drawing from the strengths of two established methods, there is leverage and it increases the rigour in a study.

In contrast to quantitative methods, the qualitative method is concerned with an understanding and interpreting of how people create and maintain their social worlds by explaining rather than predicting behaviour, and naturalistic observation rather than controlled measurement. Also, the qualitative method follows and relies on a subjective exploration of reality from the perspective of an insider as opposed to the outsider's assumptions. The contract that exists between the two methods makes them complementarily useful and significant in some IS studies.

Research design

Research design refers to the overall strategy in a study. It integrates different approaches or methods of the study into a coherent and logical way. The design is determined by the research problem and objectives. The research design provides a plan that guides the selection of critical aspects, such as the locations from which data should be collected, data collection techniques, and the units of analysis. Thus, the research design is applied to accomplish the aims of the research. It is influenced and must align with the strategy that has been adopted, as described above. The design approaches include ethnography, survey, action research, and the case study. It includes how the data was collected, analysed, and interpreted. The philosophical assumptions underlying the research are informed by the interpretivist tradition and approach. This implies a subjective epistemology and the ontological belief that reality is socially constructed.

Figure 2.1 illustrates the different types of designs that can be employed in a qualitative study. The designs might not be combined or complementarily used in a single study. However, a study may be split, depending on the size, and in such a case, two different designs might be employed. The designs are discussed, but not in holistic detail. The remainder of this section discusses the designs as covered in Figure 2.1: grounded theory, ethnography, case study, and survey. Grounded theory

Grounded theory is a design that focuses on discovering or constructing theory from data. In grounded theory, the data are systematically obtained and analysed, often in combination with comparative analysis. According to Creswell (2009), the design has two characteristics: constant comparison of data with emergence and theoretical sampling of different groups of data. In the process, it is inherently flexible, and it evolves within the context of the phenomenon being studied.

Grounded theory is a qualitative research method that seeks to develop theory from a set of grounded data systematically gathered and analysed (Urquhart, Lehmann & Myers, 2010: 357). According to Charmaz (2014), the method begins with inductive data and it involves iterative strategies of going back and forth between data and analysis. Using the grounded theory, the researcher does not begin the research process on assumptions but rather the main focus is on the empirical data gathered. Within the same line of argument, Dunne (2011) stated that the researcher is not focused on testing hypotheses taken from existing theoretical frameworks, but rather develops a new theory that is grounded in empirical data which was collected in the field.

Action research

Action Research is a systematic enquiry that focuses on a real or practical situation. It is used to develop actions and interpret the consequences through which knowledge is created. Also, it can be used to facilitate change; thus, it focuses on solving a specific problem. Although it can be used in quantitative research, it is usually applied in qualitative studies. It requires experimentation in which participants in real-life situations collaborate with the action researcher to identify and solve problems of social practice. It usually continues over an extended period of time. Baskerville and Myers (2004) argued that the primary aim of action research is to solve current practical problems while expanding scientific knowledge.

Usually, the approach has as its objective the testing of some theory or hypothesis. In the context of action research, Huang (2010) argues that theory without practice is not theory but speculation. The action researcher is not an independent observer but an active participant, and the process of change becomes the subject of research. The active participatory nature of the researcher makes the data more qualitative in nature. Thus, action research is an approach that does bear resemblance to and is often associated with qualitative research methods in that both are richly contextualised in practice.

Ethnography

This is a form of research in which the researcher (the ethnographer) participates overtly or covertly in people's daily lives for an extended time. Along the same argument, Dey (2002) empirically affirmed that during the long period of the study, the researcher's experience, in terms of his or her participation and/or observation at the research site, is used to generate a narrative-based interpretation of the events that took place. In ethnography studies, researchers are expected to spend a long period of time in the fieldwork, to ensure detailed and satisfactory observational evidence are gathered (Yin, 2015).

Using the ethnography method, data are collected through different techniques, such as questioning, observing, and listening. The use of the observation technique enables ethnographers to immerse themselves within the setting, thereby generating a rich understanding of social action (Reeves, Kuper & Hodges, 2008). However, ethnographic studies are challenged, which influences the end result of the research. According to Gilmore (2002), the ethnographic method faces some challenges beyond those of usability testing, partly because usability usually occurs later and has a more immediate connection to product quality.

Case study

The case study research method is defined as an empirical enquiry that investigates a contemporary phenomenon within its real-life context when the boundaries between phenomenon and context are not clearly evident (Yin, 2009). It offers an opportunity to formulate a detailed description of the social or organisational aspects of the social-technical interplay. Case study is particularly good for examining 'why' as well as 'how' and 'what' questions that enquire about a contemporary set of events over which the researcher has little or no control. According to Yin (2009), researchers generally do case studies for one of three purposes: to produce detailed descriptions of a phenomenon, to develop possible explanations of it, and, finally, to evaluate the phenomenon.

A case study may be either a single-case or a multiple-case design (Olivier, 2009). In a single-case study, the researcher focuses on an issue or concern and then selects one bounded case to illustrate this issue. Multiple-case designs must follow a replication rather than sampling logic in which the enquirer replicates the procedures for each case. In either single- or multiple-case designs, the case study can be embedded as well as holistic. An embedded case study is one in which there is more than one sub-unit, whilst in a holistic case study, a global programme of organisation is investigated (Yin, 2009).

As briefly discussed above, some of the design methods, such as the case study and ethnography, are similar. Myers (1999) emphasised that the main difference between case study research and ethnographic research is the extent to which the researchers immerse themselves in the life of the social group under study.

Survey

The survey design is a systematic approach used for collecting data from individuals and groups. Through a large population, it is intended to maximise insights into various topics of interest. The survey research approach is commonly used in the positivist tradition. A common goal of survey research is to collect data which is representative of a population group or groups (Kotrlik & Higgins, 2001). Questionnaires and interviews are the main techniques that are used in the data collection. In the analysis, the focus is often to draw a conclusion about relationships between particular sets of data.

However, a survey research could also be exploratory and in the interpretivist tradition, as long as the data yielded are not regarded as 'objective facts' but rather as products of the respondents' interpretation of their situation. Pfleeger and Kitchenham (2001) argued that the survey research approach can be used in a descriptive study, in which participants are requested to give their subjective view about a subject. In Lethbridge (1998), in which the survey research approach was employed, questions such as 'What is your current knowledge about this, considering what you have learned on the job as well as forgotten? How useful has this specific material been to you in your career?' were formulated and used in the data collection.

Data collection

Data collection involves the use of different techniques and approaches. Also, it is a series of interrelated activities aimed at gathering relevant information to answer the research questions that will arise or that were predefined (Cresswell, 2007). The techniques and approaches are selected based on the objectives of the study, and in alignment with the research design that are discussed above. Some of the techniques and approaches are interviews, questionnaires, observation, and document analysis.

Data sources

Oftentimes, data are collected from different sources that are usually categorised into primary and secondary. The different types of data collection techniques that are mostly associated with qualitative studies in IS are shown in Figure 2.2. The use of different methods for data collection enables the researcher to elicit different versions and forms of realities from participants. The possibility of using more than one of the techniques for data collection was suggested by Gillham (2000). In qualitative research, the primary data source is usually the interview technique, and the data from document analysis is considered secondary. The secondary data are usually existing, and may come as sources such as academics' databases, and companies' repositories, including websites. The types of materials (data) include peer-reviewed articles, white papers, strategy documents, financial reports, and annual reports. Secondary data provide essential preparation for the interviews and confirmation of some of the data gathered during and after the interviews.

The primary data source is essentially prioritised. According to Babbie and Mouton (2001), the primary data source is the data collected by the

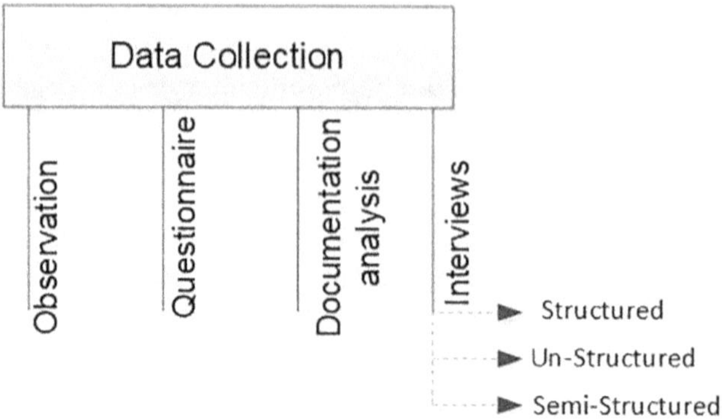

Figure 2.2 Data collection approaches.
Source: Author.

researcher (e.g., transcripts of the interviews). It is critical to master the art of collecting primary data because of the instantaneous nature of the process. Also, it is because of the clear interrelationship between the primary and secondary data. Blumberg, Cooper, and Schindler (2005) argued that primary data is required because of the proximity to reliability or truthfulness it provides, whilst the secondary data events or recordings tend to be interpreted before use or considered as data. Secondary data helps to cross-check formal information and learn about major events, technical details, and historical decisions, including organisational structure, positions, responsibilities, and roles.

Documentation

Documentation, often referred to as document analysis, can be in either hard or electronic copy. In using the document analysis technique for data collection, the researcher seeks access to the organisations' documents relating to the study. The researcher can also gather the documents or materials from sources that are outside of the organisation for which they do not need permission, such as newspaper and websites. It constitutes existing documents (materials) that have been or need to be analysed. The analysis is often not intensive; its purpose is to extract documents that are related to the phenomenon being studied.

The documents can be in the form of organisations' reports, newsletters, strategy documents, operational documents, and marketing material. The documents are intended to corroborate the evidence from other sources, such as interviews. Another important advantage of document analysis is that it may provide information that organisation members (participants) may have partly or completely forgotten in the interview process. The document analysis is a technique that is commonly used in qualitative studies in IS.

Observation

Observation is primarily concerned with studying and understanding how a group of individuals behave in their natural environment (Baker, 2006). One of the strengths of direct observation is that it enables the researcher to understand the participants' interpersonal behaviour and intentions (Paré, 2004).

Observation can be used as a powerful check against what people report about themselves. Thus, observation can be used as a complementary data collection method in an interview (Dubé & Paré, 2003). When asking behavioural questions related to the participant's past and current experiences, it is crucial that the researcher probe the responses as much as possible, until a dead-end, so as to gain a deeper understanding. In this way, the researcher is able to discover the things that participants like and dislike about their jobs, environments, and the situations that they find themselves.

Interviews

The technique involves verbal questioning of either individuals or groups. Bless, Higson-Smith, and Kagee (2006) describe an interview as a technique that offers direct personal contact with the participants who are requested to answer questions related to the research problem. Interviews can be conducted in a formal or in an informal environment, depending on the suitability of both the interviewer and the interviewee.

In Yin's (2009) explanation, interviews are primarily to elicit information based on views, opinions, and experiences from respondents, but in their own words, rather than to collect data that is simply a choice among pre-established response categories. One of the advantages of interviews includes good cooperation from respondents, which makes it possible to reach all respondents, both literate and illiterate (Cooper & Schindler, 2003). It also makes it possible for the researcher to reach areas of reality that would otherwise remain inaccessible, such as people's subjective experiences and attitudes (Denzin & Lincoln, 2005). There are different types of interviews, namely, structured, semi-structured, and unstructured interviews.

Structured interview

In the structured interview approach, structure is provided to guide the interviewer. The structured interviews are fixed, which means that the questions are predefined or prepared in advance. The questions are asked sequentially in the same order to each interviewee. The technique is sometimes referred to as directive interview. Although this style lacks the free flow of a friendly conversation like an unstructured interview, it provides the correctness and reliability needed in some situations. Structured interviews ensure candidates have equal opportunities to provide information and are assessed accurately and consistently. It is applied in quantitative studies.

Unstructured interview

In unstructured interviews, there are no predefined questions. The interview questions are changed or adapted to meet the respondent's acumen, understanding, or belief (Kvale & Brinkman, 2009). Unlike the structured interview technique, it does not offer a restricted, pre-set range of answers for a respondent to choose from, but instead advocates listening to how each individual person responds to the question. The unstructured interviews are considered as a way to understand complex behaviour of people without imposing any priori categorisation which might limit the field of enquiry. The technique has a bit of the semi-structured interview, from the perspective of instant probing of respondents' responses. The technique is usually used in qualitative studies.

Semi-structured interview

Semi-structured interviews are considered to be a non-standardised form of data collection. The technique is frequently used in qualitative research. The semi-structured interview provides more insights and scope for discussion and learning about the phenomenon being studied. It includes individuals' opinions, experiences, and views. In the semi-structured interview technique, the order of questions can be changed, depending on the direction of the interview process. It, however, gives the researcher opportunities to probe the interviewee for views and opinions. Gray (2004) argued that probing is employed to explore new paths that were not initially considered. Probing is used to clarify and uncover answers where participants were required to explain or build on their responses. According to Corbin and Strauss (2008), this is intended to provide rich data for analyses and interpretations. The technique is mostly associated with qualitative studies.

Consent is requested and granted before an interview can be carried out with an individual or a group. A guide is normally used in both structured and semi-structured interviews. An interview guide contains specific issues and questions to be considered when commencing the interview. The main function of an interview guide is to serve as a checklist for the interviewer to ensure that the same issues are addressed in every interview and not forgotten in some interviews (Blumberg, Cooper & Schindler, 2008: 387).

In qualitative studies, it is tricky to know when to stop the process of interviewing. According to Yin (2003), there is no limitation to the number of interviews that are conducted in qualitative studies. However, the process has to stop at some point, when the point of saturation is reached, which means that there is no new information coming forth from the interviewees. Thus, when themes and issues in which the researcher is interested become saturated, the process of data collection should come to an end.

Data analysis

Data analysis is a practice in which raw data is ordered and organised so that useful information can be extracted from it (Holloway, 1997). The process of organising and thinking about data is important to understanding what the data does and does not contain, in terms of meaning and explanation of the phenomena being studied.

The different theoretical concepts, such as activity theory, actor-network theory, contingency theory, diffusion of innovation theory, and structuration theory, emphasise different social contexts and therefore facilitate different types of meanings and explanations. However, each of the theories does bring fresh perspectives into IS studies. The theories enable examining the interaction and relationship that exist between actors, people, and technology, including the networks which they form in a conscious or unconscious manner, in the development, implementation, use, and management of IT solutions. Thus, the interpretivist approach without sociotechnical theory can by itself be short in its sole use for data analysis in IS studies. Chapter 1 introduced some of the theories which are extensively discussed from Chapter 4 to 12.

Conclusion

In this chapter, the qualitative interpretive research approach, including the rationale for employing the methods, is introduced, and discussed. The research design and methods for addressing data collection and analysis are also briefly discussed. In addition, the theoretical and philosophical assumptions underlying research methodology in the IS field are reviewed, but not in detail. The chapter provides a summary of the methods, techniques, and approaches that can be employed in collecting data in a qualitative study.

The analysis aspect focuses on people's subjectivity as they continue to make use of technology for different reasons and in various places. Thus, the stability or continuous change of IT solutions is subject to human actions and interactions. Hence, examining human actions and the manifestation of their interactions is critical, using different sociotechnical theories. As discussed in the remainder of the book, theoretical lenses are intended to assist researchers to gain a deeper understanding of the phenomena being studied.

References

Archer, M. (1995). *Realist social theory: The morphogenetic approach.* Cambridge: Cambridge University Press.

Babbie, E., & Mouton, J. (2001). *The practice of social research.* Cape Town: Oxford University Press.

Babbie, E. (2007). *The practice of social research.* Belmont: Thomson.

Baker, L. M. (2006). Observation: A complex research method. *Library Trend, 55*(1), 171–189.

Baskerville, R., & Myers, M. D. (2004). Special issue on action research in information systems: Making IS research relevant to practice foreword. *Management Information Systems Quarterly, 28*(3), 329–335.

Bless, C., Higson-Smith, C., & Kagee, A. (2006). *Fundamentals of social research methods: An African perspective.* Cape Town: Juta.

Blumberg, B., Cooper, D., & Schindler, P. (2005). *Business research methods.* London: McGraw Hill.

Blumberg, B., Cooper, R. D., & Schindler, P. S. (2008). *Business research methods* (2nd ed.). Berkshire: McGraw-Hill.

Cecez-Kecmanovic, D. (2005). Basic assumptions of the critical research perspectives in information systems. In: D. Howcroft & E.M. Trauth (eds.), *Handbook of critical information systems research: Theory and application*, pp. 19–46. Cheltenham: Edward Elgar.

Charmaz, K. (2014). *Constructing grounded theory* (2nd ed.). London: SAGE Publications Ltd.

Chen, W., & Hirschheim, R. (2004). A paradigmatic and methodological examination of information systems research from 1991 to 2001. *Information Systems Journal, 14*(3), 197–235.

Cooper, R. D., & Schindler, P. S. (2003). *Business Research Methods. International ed.* New York: McGraw-Hill/Irwin.

Corbin, J., & Strauss, A. (2008). *Basics of qualitative research (3rd ed.): Techniques and procedures for developing grounded theory.* London: SAGE Publications, Inc.

Cresswell, J. W. (2007). *Qualitative inquiry & research: Choosing among Five approaches.* Thousand Oaks, CA: Sage Publications.

Creswell, J. W. (2009). *Research designs: Qualitative, quantitative and mixed methods approaches* (3rd ed.). Thousand Oaks, CA: Sage Publication.

De Vos, A. S., Strydom, H., Fouche, C. B., & Delport, C. S. L. (2002). *Research at Grass Roots: For the social sciences and human service professions.* Pretoria: Van Schaik.

Denzin, N. K., & Lincoln, Y. S. (2005). *The SAGE handbook of qualitative research* (3rd ed.). London, Thousand Oaks, CA: Sage Publications.

Dey, C. (2002). Methodological issues: The use of critical ethnography as an active research methodology. *Accounting, Auditing & Accountability Journal, 15*(1), 106–121.

Dobson, P. J. (2002). Critical realism and information systems research: Why bother with philosophy. *Information Research, 7*(2), 7–2.

Dubé, L., & Paré, G. (2003). Rigor in information systems positivist case research: Current practices, trends, and recommendations. *MIS Quarterly, 27*(4), 597–636.

Dunne, C. (2011). The place of the literature review in grounded theory research. *International Journal of Social Research Methodology, 14*(2), 111–124.

Galliers, R. D. (1991). Strategic information systems planning: Myths, reality and guidelines for successful implementation. *European Journal of Information Systems*, 1(1), 55–64.
Gillham, B. (2000). *The research interview.* New York: Continuum.
Gilmore, D. (2002). Understanding and overcoming resistance to ethnographic design research. *Interactions*, *9*(3), 29–35.
Goldkuhl, G. (2012). Pragmatism vs interpretivism in qualitative information systems research. *European Journal of Information Systems*, *21*(2), 135–146.
Gray, D. E. (2004). *Doing research in the real world.* London: Sage Publication.
Hassan, N. R., Mingers, J., & Stahl, B. (2018). Philosophy and information systems: Where are we and where should we go? *European Journal of Information Systems*, *27*(3), 263–277.
Hirschheim, R., & Klein, H. K. (2012). A glorious and not-so-short history of the information systems field. *Journal of the Association for Information Systems*, *13*(4), 188–235.
Hirschheim, R., & Klein, H. K. (1989). Four paradigms of information systems development. *Communications of the ACM*, *32*(10), 1199–1216.
Holloway, I. (1997). *Basic concepts for qualitative research.* London: Blackwell Science.
Huang, H. B. (2010). What is good action research. *Action Research*, *8*(1), 93–109.
Kelliher, F. (2005). Interpretivism and the pursuit of research legitimisation: An integrated approach to single case design. *The Electronic Journal of Business Research Methodology*, *1*(2), 123–131.
Kelly, M. (2009). The role of theory in qualitative health research. *Family Practice*, *27*(3), 285–290.
Klein, H. K., & Myers, M. D. (1999). A set of principles for conducting and evaluating interpretive field studies in information systems. *MIS Quarterly*, *23*(1), 67–93.
Kock, N., Avison, D., & Malaurent, J. (2017). Positivist information systems action research: Methodological issues. *Journal of Management Information Systems*, *34*(3), 754–767.
Kotrlik, J. W. K. J. W., & Higgins, C. C. H. C. C. (2001), Organizational research: Determining appropriate sample size in survey research appropriate sample size in survey research. *Information Technology, Learning, and Performance Journal*, *19*(1), 43–50.
Kvale, S., & Brinkmann, S. (2009). *Interviews: Learning the craft of qualitative research interviewing* (2nd ed.). Thousand Oaks, CA: Sage.
Leedy, P. D., & Ormrod, J. E. (2010). What is research. *Practical research planning and design* (9th ed.), pp. 1–11. Upper Saddle River, NJ: Prentice Hall.
Lethbridge, T. C. (1998). A survey of the relevance of computer science and software engineering education. In Software Engineering Education, 1998. In Proceedings of 11th Conference on (pp. 56–66). IEEE.
Mingers, J., Mutch, A., & Willcocks, L. (2013). Critical realism in information systems research. *MIS Quarterly*, *37*(3), 795–802.
Mouton, J. (2001). *How to succeed in your master's and doctoral studies: A South African guide and resource book.* Pretoria: Van Schaik.
Myers, M. D. (1997). Qualitative research in information systems. *Management Information Systems Quarterly*, *21*(2), 241–242.
Nenty, H. J. (2009). Writing a quantitative research thesis. *International Journal of Science Education*, *1*(1), 19–34.

Olivier, M. S. (2009). *Information technology research: A practical guide for Computer Science and Informatics* (3rd ed.). Pretoria: Van Schaik

Orlikowski, W. J., & Baroudi, J. J. (1991). Studying information technology in organizations: Research approaches and assumptions. *Information Systems Research*, *2*(1), 1–28.

Paré, G. (2004). Investigating information systems with positivist case research. *Communications of the Association for Information Systems*, *13*(1), 233–264.

Pfleeger, S. L., & Kitchenham, B. A. (2001). Principles of survey research: Part 1: Turning lemons into lemonade. *ACM SIGSOFT Software Engineering Notes*, *26*(6), 16–18.

Reeves, S., Kuper, A., & Hodges, B. D. (2008). Qualitative research methodologies: Ethnography. *BMJ*, 337(aug07_3), a1020-a1020.

Rowe, F. (2018). Being critical is good, but better with philosophy! From digital transformation and values to the future of IS research. *European Journal of Information Systems*, *27*(3), 380–393.

Sekaran, U. (2003). *Research methods for business* (4th ed.). New York: John Wiley & Sons, Inc.

Smith, M. L. (2006). Overcoming theory-practice inconsistencies: Critical realism and information systems research. *Information and Organization*, *16*(3), 191–211.

Straub, D., Boudreau, M. C., & Gefen, D. (2004). Validation guidelines for IS positivist research. *Communications of the Association for Information Systems*, *13*(1), 380–427.

Trauth, E. M., & Jessup, L. M. (2000). Understanding computer-mediated discussions: Positivist and interpretive analyses of group support system use. *MIS Quarterly*, *24*(1), 43–79.

Urquhart, C., Lehmann, H., & Myers, M. (2010). Putting the 'theory' back into grounded theory: Guidelines for grounded theory studies in information systems. *Information Systems Journal*, *20*(1), 357–381.

Venkatesh, V., Brown, S. A., & Sullivan, Y. W. (2016). Guidelines for conducting mixed-methods research: An extension and illustration. *Journal of the Association for Information Systems*, *17*(7), 436–476.

Walsham, G. (1993). *Interpreting information systems in organizations* (Vol. 19). Chichester: Wiley.

Walsham, G. (1995). The emergence of interpretivism in IS research. *Information Systems Research*, *6*(4), 376–394.

Yin, R. K. (2009). *Case study research: Design and methods* (4th ed.). Thousand Oaks, CA: Sage.

Yin, R. K. (2015). *Qualitative research from start to finish*. New York: Guilford Publications.

Zachariadis, M., Scott, S., & Barrett, M. (2013). Methodological implications of critical realism for mixed-methods research. *MIS Quarterly*, *37*(3), 855–879.

3 Qualitative data for information systems research

Table of contents

Introduction

In the last two decades, information systems (IS) researchers have increasingly employed an interpretivist approach and qualitative methods in their studies (Stokes & Urquhart, 2013). Within the qualitative methods, many IS researchers, including postgraduate students, continue to apply the semi-structured interview even though there are challenges with the technique, notably because it is incapable of generating huge data. Perhaps increasing use of the technique is based on the fact that many of the challenges have in the past been highlighted and addressed in studies, such as Myers and Newman (2007) and Schultze and Avital (2011). Of course, not all the challenges were addressed, and they continue to have an impact on IS aspiring researchers and postgraduate students using the semi-structured interview in their research pursuits. This is either because they are not aware of the pitfalls or because they take them for granted.

Qualitative study's reliance and increasing popularity may be associated with the power to warrant successes in generalisation (Katz, 2015). This premise is fortified by the approaches and techniques in the collection and analysis of data. Qualitative methods guide a study to explore, and, more importantly, to obtain a depth of understanding as to why there is success or failure in the phenomenon being studied. Such an understanding is essential because of the complexity which human actors bring into IT solutions. Thus,

DOI: 10.4324/9781003184119-3

intensive interviews are required to gain an understanding and knowledge of what, how, and why things happen in the way that they do, whether in the development, implementation, or maintenance of IT solutions.

Semi-structured interviews are particularly suitable for collecting qualitative data mainly because they provide the opportunity for exhaustive discussion, through which complexities are explained and new topics emerge. Creswell and Poth (2017) argued that the main challenge in using qualitative methods is that the interviewer often focuses on the mechanics of conducting the interviews. In such circumstances, the interviewer has a responsibility to motivate the participants to feel at ease, in order to collect comprehensive and quality data. That is one of the strengths of the semi-structured interview technique; it allows negotiation between the interviewer and interviewees.

Comprehensiveness in the process of data collection helps to ensure that key aspects of the phenomenon are not missed. A comprehensive approach entails holistic inclusiveness of participants and detailed interaction with each of them. Also, flexibility, which allows participants to freely express themselves during the process of data collection, significantly contributes and ensures relevance and richness of data. Cleary, Horsfall, and Hayter (2014) explained that the extensiveness of data collection processes, from the perspectives of breadth, depth, and scope, is of critical importance to qualitative studies. This is a measure for addressing complexity in IS studies. Qualitative research often focuses on investigating and understanding complex phenomena from social and cultural perspectives to technical angles. Thus, Avison and Malaurent (2014) suggested that using the interpretivist approach, it is essential to preserve a considerable degree of interaction during data collection, in investigating factors that influence and determine the phenomena being studied.

The primary focus of this chapter is to examine the challenges that are encountered by some researchers in the process of collecting qualitative data, using the semi-structured interview technique. Thus, the question that the chapter tries to address is, what are the factors that influence data collection, using the semi-structured technique? This is to help with an understanding of why sociotechnical theories should be applied in some qualitative studies. The remainder of this chapter is organised into five main sections as follows: The first section presents a discussion about qualitative methods and semi-structured technique from literature perspective. The data collection process model for interviews is presented in the section that follows. The third section presents a more specific model for the semi-structured interview technique in qualitative studies. The implication of practice is covered in the fourth section. The discussion includes the challenges that are considered to be critical and have a high likelihood of negative impact on quality data in IS studies. The conclusion is drawn in the fifth section.

Qualitative methods and semi-structured interview

In the last three decades, the use of qualitative methods and the semi-structured technique in IS studies has increased tremendously (Bélanger & Crossler, 2011; Hussain, Wallace & Cornelius, 2007). Both the qualitative methods and the semi-structured technique are often associated with the interpretive approach (Walsham, 2006). The increasing use of the methods and the technique can be attributed to their benefits. In Myers' (2013) view, some of the key benefits of qualitative research are that it allows researchers to see and understand the context within which decisions and actions take place, over a period of time.

Qualitative methods are often employed in scientific research that systematically make use of a predefined set of procedures to seek answers to questions, collect evidence, and produce findings that are applicable beyond the immediate boundaries of the study (Denzin & Lincoln, 2000). Myers (2009) describes the qualitative method as a research method that enables researchers to study phenomena in their own natural settings, which is essential in the real world. Qualitative research methods are more about the social world with a specific focus on cultures and ways of life, rather than about seeking causal explanations for cultural practices (Silverman, 2013). Thus, the methods are particularly used to study individuals' subjective views and opinions about their environment, natural settings, and associated components. Therefore, it is near-impossible to understand why and how things happen in the way that they do in organisations without any interaction with the actors (Bryman, 2012).

Despite the numerous benefits, there are many challenges with qualitative methods. Thus, Baškarada (2014) cautions against the type and how data is collected, as it must be logically linked to the research conclusions. According to Schultze and Avital (2011), in gathering qualitative data, some respondents experience difficulties in attempts to envision futuristic and ideal worlds, in that their imagination is constrained by their view of the past and the present. Another challenge that has been identified is that it is difficult to listen to an interviewee and construct a follow-up question for further probing of responses at the same time (Myers & Newman, 2007).

Through the semi-structured interview technique, data collection is usually interactive and enables researchers to learn how individuals and groups act or respond to different circumstances in their environments (Tsang, 2014). Katz (2015) suggests that interactions between interviewers and interviewees substantially improve explanations in the phenomena being studied. Polak and Green (2016) argued that the interview technique is an appropriate approach for studying complex phenomena and shared practices, in that it promotes exploration of detailed data through interaction. The semi-structured interview is flexible and allows an extended interaction and a variety of responses, which sometimes reduce the risk of bias.

However, there are many challenges with gathering qualitative data using the semi-structured interview technique: (1) Avison and Malaurent (2014) explained that there is no formula for collecting qualitative data in empirical studies, making it more complex than what it actually is; (2) even though Cleary et al. (2014) argued that qualitative studies rely on interactive interview-based spoken language, clarity, and explicitness in gathering in-depth data, Choy (2014) explained that the technique is problematic in that the participants have more control over the content of the data collected; and (3) interviewers struggle with capacity to generate rich data about interviewees' experiences of their real worlds, and the meanings that they associate with them (Schultze & Avital, 2011).

Many of the researchers often employ the interpretive approach in qualitative research. Goldkuhl (2012) explains that the core idea of interpretivism is to understand the subjective meaning of persons. In Walsham's (2006) summation, the interpretive approach in IS research is used to adopt our knowledge of a reality that is consciously or unconsciously constructed by human actors. Klein and Myers (1999) argued that the hermeneutics technique makes interpretive research explicit and determines reasons for the understanding of a text. Hermeneutics does not aim to explain and predict but to examine the phenomena being studied, through an understanding and sense making of the actions that take place.

Data collection process model

The semi-structured technique of the qualitative method is considered efficient in collecting rich data in IS studies. However, it depends on how the technique is applied, and this varies from one researcher to another. Also, before analysis can be conducted, it is important to know how the data were collected and understand the factors that influence the process. As explained further below, four factors primarily influence data collection in IS studies: (i) unconscious bias; (ii) data size; (iii) knowledge of the subject; and (iv) the transcription process.

Data collection is significant in research because of its importance in the evaluation and success of a study. The data collection process is directly influenced by the semi-structured interview that is used, which can be cumbersome. This section presents the data collection process model (Figure 3.1), which consists of six main attributes. The model shows the complementary use of two techniques, namely, semi-structured interviews and document analysis for data collection. The explanation that follows should be read with Figure 3.1 in mind to gain a better understanding of the process.

Interview questions

In achieving the research objectives, interview questions are extracted from the research questions. Interview questions form the written guidelines that are used to elicit information from participants. The guidelines

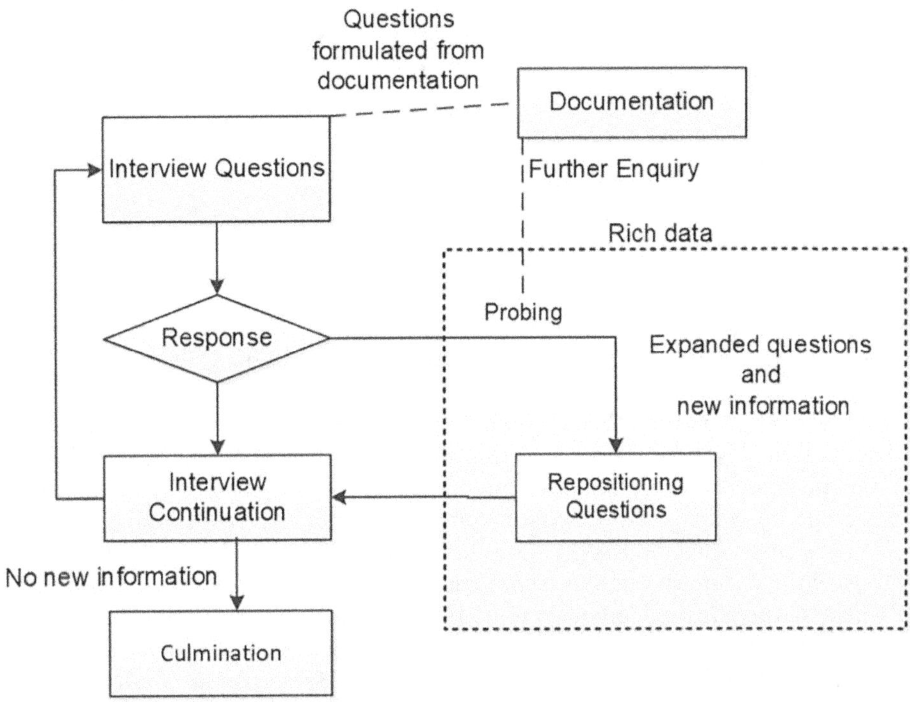

Figure 3.1 Data collection process.
Source: Author.

help the researcher to maintain consistency, uniformity, and order during the process of data collection, thereby contributing towards improving the quality of the data, without expectation or bias.

Expectation and unconscious bias – Within the qualitative methods, the interpretivist approach is employed to gain an understanding of the phenomenon by exploring the meanings which people associate with things within context. Even though the approach does not predefine the phenomenon, there is a conscious or unconscious tendency to continue to misconstrue the purpose of interview guidelines in an interpretivist approach.

The interpretive research approach focuses on sense making of activities within the context, rather than a predefinition of dependent or independent variables (Myers, 1997). The nature of the interpretive approach makes it difficult to completely separate the researcher from the phenomenon being studied. Following Klein and Myers' (1999) suggestions, it is against the principle of the interpretive approach to collect data with a particular mindset, which is bias. Thus, data that are sometimes thought to be useless can become the corner stones or focal points during the analysis.

Data size – The usefulness and relevance of qualitative data are not determined by the size or volume, but by depth and richness. In a study where the aim is not to test or validate a value or a variable, there is no need for the researcher to expect answers from the respondents. Qualitative methods are not intended to test hypotheses, but to produce an understanding within a specific setting, whereby the phenomenon being studied influences and is influenced by the same social context. Such culmination helps to determine or influence the data size and could be attributed to the researcher's knowledge of the subject.

Document analysis

Examining documents is referred to as document analysis, which is used for collecting qualitative data. The technique is used to obtain or access existing documents that are related to the phenomena being studied. The documents can be in hard or soft (electronic) copies. Examining of documents helps to provide valuable data that would have otherwise been difficult or impossible to obtain if only the interview technique was employed. Thus, the technique is considered to improve data quality. Also, data from the technique are compared and aligned with data from interviews.

The interview questions can also be generated from existing documents by using the document analysis technique. This occurs when the researcher finds items in the existing documents which directly reflect on the research objectives. Hence, it is necessary to request access to available documents that are related to the study before interviews are embarked upon. However, data obtained from documents often require further clarification. Thus, the researcher incorporates such aspects of data into the interview questions for the purpose of clarity. This process enhances data quality and richness. The documents obtained for the purpose of a study should be reviewed as soon as possible or practical, and not after the interview process has been completed. This is primarily because the documents can help the researcher to identify critical areas that might require further exploration and probing during the interview process.

Response

Responses are subjective opinions and views of interviewees when answering questions where a semi-structured technique is used. In the process of interviews, the researcher should take written notes in order to capture as many responses as possible from the interviewees. Thus, the researcher is strongly advised to listen attentively even if the interview is being recorded. By doing so, the researcher stays focused and is able to grasp as much details as possible and therefore able to identify areas that need further probing and clarifications. Also, in order to probe responses further, the researcher must be knowledgeable enough about the phenomenon being studied. The lack or limited knowledge of the subject limits the researcher's ability to explore and probe the interviewees' responses. Hence, it is critical to have done a great deal of work in the area of study, including a review of related literature before data collection is undertaken.

Knowledge of the subject – Consulting a wide range of literature helps to gain knowledge and a better understanding of the phenomenon being studied. Essentially, this helps to guide and manage the interview process, particularly when an interviewee attempts to dictate the direction. As a result, researchers are inclined to build on their knowledge of the subject that is being studied, which is necessary for a robust conversation with the respondents during data collection.

Some researchers learn more about their studies as they progress and during data collection. Others have a vast knowledge of the phenomenon before they embark on the data collection stage, which enhances their ability to probe responses. On the one hand, limited knowledge limits the researcher from taking advantage of the flexibility which the interpretive approach allows, using a semi-structured interview technique. On the other hand, extensive knowledge could make the researcher lead the participants in a specific direction or influence the participants' views. The balance between knowing and not knowing enough is a challenge, although it has long been trivialised among researchers and postgraduate students.

Probing of responses is crucial in interpretive studies, particularly when using the semi-structured interview technique. Through the probing of responses, the researchers expand on their knowledge, as well as adjust the existing interview questions. This leads to a collection of new information, which enables rich data to be obtained. In areas where probing is not done, there are likely to be gaps in the data collected. This creates a gap either during the analysis of the data or in the results that are reached. There is no limit to the number of questions that are posed during the probe of the responses. As shown in Figure 3.1, responses can be iterative as probing continues. For quality purposes, the interviewer stops probing at the point when all questionable responses have been clarified. To end the probing process, the researcher repositions the questions by going back to the interview guidelines.

Repositioning questions

The purpose of repositioning is to avoid diverting attention from the original topics during the probing of interviewees' responses. This approach allows the interviewer to reconnect with the interview questions (guidelines). By reconnecting with the interview questions, the researcher is able to maintain the flow of the enquiries towards achieving the research objectives. Repositioning becomes even more necessary when both interviewer and interviewee are passionate and very knowledgeable about the discourse; otherwise, the researcher loses track of the direction of the responses.

The implication of not reconnecting with the interview questions as quickly as possible could be costly, in that the researcher might end up collecting quantitative data for qualitative purposes. This happens when 'true or false' is offered as responses during probing. This is as a result of inconsistencies which manifest because of a lack of flow in the questions being asked. Also, this could lead to respondents getting lost or confused in the process, as the manner in which the interview questions are being posed becomes disjointed. This makes it difficult to continue with the interview from either the interviewee's or the interviewer's or both perspectives. In addition, it becomes difficult or impossible for the researcher to link the sets of data during analysis. In such a situation, the researcher is required to conduct extra fieldwork. Another point is that it is important not to delay transcribing the data for personal reasons. This is primarily because if it takes so long before the data is transcribed, the researcher might forget or not be able to reflect on some moments and incidents during the interview process accurately. Some of the moments can be critical, especially if some words are not clear on the recording device.

Data transcribing – Data can be recorded with various types of devices, such as cellular phones and electronic tapes, since the data are both texts and voices. Transcribing the recorded data can be a rigorous process on its own. This is because if the transcribing is not accurate, the researcher may lose the depth and richness of the data. As such, transcribing should be carried out with a high level of diligence and appropriateness, including the use of reliable devices for clear audio sound.

In addition, the other tasks entail organising and cleaning of the data for the analysis, which wholly depend on how well the researcher is familiar with the data sets. Also, some data get lost along the way, during the transcribing and documentation processes. This makes the role of the researcher in the management of the data critical.

Interview continuation

Continuation is the step-by-step process of data collection by using the interview technique. The process prevents a disjoint and strengthens focus during data collection. Disjoint or focus strengthening can happen, irrespective of whether probing did or did not occur in the process of the interview. However, if an interviewee's response requires further probing, continuation is done after the repositioning process is complete. If no probing is required, the researcher proceeds to the next question as stated in the interview guidelines.

The outcome of one interview is a guide to conducting the next, in that it is used for corrective measures and improvement of data for achieving the objectives of the study. Therefore, it is necessary for the researcher to listen attentively to the audio as they transcribe the data. In the process, it helps the researcher to make sense of the data and avoid discarding valuable data. The interview process can be iterative in nature, which the semi-structured technique allows. However, the process has to stop at some point. The process ceases when a point of saturation is achieved, a point where no new and unique information is forthcoming but, rather, there is a repetition of what has been said or opined by previous respondents. Point of saturation requires the researcher's familiarity with the data they have collected in the process. This can be achieved through transcribing the recorded interviews immediately after each interview has been conducted. Thus, accumulating the interview materials is not a good practice and it limits the researcher's chances of gathering rich data. This is primarily because the researcher will not be familiar with the data already collected in order to identify where gaps might exist.

Culmination

Culmination refers to a process whereby data collection comes to an end in a study. Irrespective of technique or size, this has to happen at some point during any study. The end point can be initiated either by the researcher or by the interviewee. The researcher often ends the process at the point where the questions in the interview guidelines have been exhausted. The interviewee ends the process when the response leads to closure. This is rare.

Thus, culmination is considered a dead-end in discussion between the researcher and the participants on matters relating to the phenomena being studied. Thereafter, the researcher performs data cleansing in preparation for analysis. This is done by proofreading and editing, as well as appropriately labelling the texts from the interviews. Without reaching the point of culmination, it is difficult to avoid disjointedness and inconsistency in the data, which determines and influences the findings.

Semi-structured interview technique for qualitative data

The aim is to examine and understand the factors that influence the collection of qualitative data, using the semi-structured technique. This entails data collection, including the conduct of both the interviewer and interviewees. Also, the discourse draws on the role of the environment in which the data is collected. There are five factors that fundamentally influence data collection in IS studies: (1) face-to-face interview; (2) telephonic approaches; (3) tape recording an interview; (4) guiding principles to data collection; and (5) demographics of participants. The influence of each of these factors is either direct or indirect.

Face-to-face interview

The face-to-face approach traditionally uses the semi-structured technique. It can be defined as 'conversational', in that the 'interviewer establishes a general direction for the conversation and pursues specific topics which are raised by the participant' (Babbie, 2007: 306). The approach allows both the researchers and the participants to have direct face-to-face (physical presence) contact and conversation. Many postgraduate students employ the face-to-face approach mainly because they have to put the faces of the participants to their voices and personalities, which helps to decide whether to be formal, informal, or semi-informal in the process. Also, the way interviewers present themselves significantly influences how interviewees respond, whether to be more open and have an extensive conversation or shy away during the interviews. The semi-structured interview technique allows interviewees to express themselves as much as possible, enabling interviewers to see things from many angles. It can therefore be regarded as a good data collection method to obtain or gain an in-depth understanding of the participants' experiences, viewpoints, and opinions on the phenomenon being studied.

The face-to-face approach promotes richness of data through a comprehensive understanding of participants' views and opinions, including the opportunity to instantaneously probe responses as much as possible. Also, the approach provides a chance to make sense of the respondents' body language during the interviews and the environments where the process takes place. In order to create a friendly environment, some interviewers start with few interactions and jokes with the participants before the commencement of the interview. Consequently, it reduces tense situations and allows questions as many times as possible. Particularly to postgraduates, some of the interviewees can be very intimidating or cold in the beginning of the interview process. But as the interview progresses, some interviewees warm up to the interviewers, fortifying stable moments and increasing levels of confidence.

Due to the unpredictable nature of some environments, some interviewers often have different experiences when it comes to culture and tradition. As a result, some interviewers encounter various reproaches in areas where they collect data. This is because they try to apply the same traditional approach across the different environments where they collect data. Each societal or company environment is unique. For example, there are certain ways in which elders in some communities would like to be greeted or approached, which a researcher or interviewer might not be privy to. As a result, both the interviewer and the interviewee are embroiled in cultural conflict.

Another challenge is spoken language and accents. Even when both the interviewer and interviewees speak the same language, there is no guarantee of fluency and clear understanding of each other because of different accents. Words are pronounced differently, and this affects the understanding of questions or responses. In some cases, either the interviewer or the interviewee is compelled to spell out or literally write down the words or sentence. It can be either fun or a derailment of the process, depending on the character of the interviewee. This makes the area or venue where the data is collected important. At the point where accents become an obstructive element, the interviewer or interviewee can carry out illustration for the purposes of clarity and better understanding. Such challenges consequently cause collection of inappropriate, wrong, or less rich data, without the face-to-face approach.

In addition, the face-to-face approach makes it quicker and easier for the interviewers and interviewees to be familiar and freer with each other. Once the interviewers and interviewees are free or familiar with each other, it becomes easier for them to express themselves as much as they possibly can. On the one hand, it gives the participants the opportunity to ask for clarity if and where they do not understand a question. The familiarity leads to inquisitiveness and results in more questions and answers, thereby enriching the data. On the other hand, it allows the researcher to instantly ask follow-up questions or get clarification if and when the need arises. A mix of formal and informal terminologies can be helpful in getting a lot out of the interviewees. This is only possible if friendliness is established before the interview process begins, making both the interviewer and the interviewee relaxed with each other.

One of the advantages of using the technique is the relationship that is created between an interviewer and interviewees during the interview process. In some cases, the relationship extends beyond the period of interview or the study. The relationship is an affirmation of the social inclination that is involved in qualitative studies. Also, the face-to-face process is a controlled interview situation where the interviewer has the possibility to ask complicated questions, as well as follow up on the questions. Another important factor is that the technique is socially focused; therefore, it is even more helpful when human perceptions and behaviours are part of the phenomenon being studied. The technique allows the interviewer to gather in-depth data about the topic, through continuous probing of responses, until they feel it has been fully exhausted, as the participants express their thoughts and feelings (Katz, 2015). Sekaran (2003) emphasises the fact that the semi-structured interview technique is flexible, allowing new questions to be generated during an interview, based on interviewees' responses.

Telephonic interview

The choice of telephonic approach for interviews can be based on many factors. Three of the factors that are essentially decisive are distance, availability, and the contagious nature of pandemic diseases. In such circumstances, the interviewer does not have control or choice but has to rely on or employ the telephonic interview approach. The decision to apply the telephonic approach is not always made by the interviewer alone; the interviewee can also be given the option.

The distance between the researcher's or postgraduate student's residence or location and their target interviewees is always a major determinant in that many projects have budget constraints. This gets worse when there are many participants to be interviewed and they are sparsely spread, requiring more travelling, which can be prohibitive. Thus, the use of the telephonic approach becomes economical. Facelessness is another rationale why some researchers, particularly aspiring researchers, or postgraduate students, prefer the telephonic approach. The approach allows the interviewer to hide behind facelessness and be in command instead of being in front of executive personnel that intimidate. In addition, the availability of participants is always problematic. Many interviewers often complain about cancellation or postponement of interview appointments. Thus, the telephonic approach allows the interviewee to multitask (e.g., travelling) while the interview is ongoing.

The challenges that are caused by the unavailability or lack of physical presence of the participants makes the telephonic interview approach a viable option. The approach can be convenient to the participants, in that search and preparation of interview meeting venues are avoided. This means time and resources are saved. However, there are some factors that can add to the challenges of integrity and quality of the data that are gathered using the telephonic interview approach. For example, since it is a faceless engagement, the interviewer loses out on the opportunity to analyse and interpret the environment and body language of the interviewee, which can be used to complement some responses. Also, some interviewees are particular on the manner of interaction, based on their cultural orientation and affiliation. In instances where such manner is not adhered to, responses can be reduced to minimal. A researcher once related an experience where he mistakenly referred to a lady interviewee as a male many times during an interview. *She got irritated and dropped the call. I did later find out that she is indeed a female*, the researcher said.

However, the approach can be used complementarily with the face-to-face approach. In such a harmonising approach, the telephonic approach can be used as a follow-up after a face-to-face interview has been conducted. It makes the process easier for both the interviewer and the interviewee. It gives the former time to reflect on the data gathered and the interviewer may thereafter ring the latter for possible clarification or further expansion. Importantly, regardless of the approach that is selected and used, the same guidelines are employed. This is important for the consistency of the interview process.

Tape recording an interview

It has become a common practice for many IS researchers to tape record their interviews. One of the justifications for employing this approach is that it allows the researcher to listen more attentively and have enough time to reflect, assess, and articulate follow-up questions from interviewees' responses. However, the use of a tape recorder makes some interviewees uncomfortable. Hence, some participants do not allow the use of a tape recorder without permission. Also, the permission to allow tape recording depends on the code of conduct that guides the research and the policy of the organisation that is used as a case in the study. This could be why *some interviewees request for the transcript of the interview, to validate what has been recorded.* This could be attributed to the fact that many people who participate in academic research are not aware that it is an innocent request to tape record the conversation. Hence, a proper explanation is required before the commencement of every interview process.

The request to tape record interviews is often because not all researchers are able to capture interviewees' responses perfectly and accurately during the interview process, especially when some participants speak fast. Without the use of a tape recorder, the process poses more challenges for first timers in collecting qualitative data, using the semi-structured interview approach. In theory, it is easy and straightforward to include the recording of interviews in the interview guidelines and procedure. In practice, it is a bit complex, in that it depends on factors such as individual mood and cultural belief, as well as the ethics of the environment (or organisation) where the interviews are conducted.

Thus, the intention to tape record an interview conversation must be negotiated and clarified between the researcher and the interviewee because of its significance. The use of a tape recorder prevents the interviewer from losing focus and track of the interview process, as note taking could cause distraction. The use of a tape recorder for interview sessions allows the interviewer to aggressively listen to the interviewee. This can sometimes encourage and motivate some interviewees to share more information about the phenomena being studied. After the interviews, the researcher repeatedly listens attentively to the recorded conversations. Cognitively, it is helpful for accurate situational reflection and comparison with the interviewer's notes when transcribing the data.

The duration for which the interview lasts is not always important, as long as the quality and richness of the data remain the absolute focus. However, it should be documented in the study by way of adding credibility to the data collection process. Also, *the tape accurately captures the time spent in the interview process and this adds value from the perspective of documentation of the entire process of data collection.* This is to improve on the quality of the research statistical procedure and can also be used to problematise further research.

Each of the approaches – face-to-face, telephonic, and tape recording – should be guided by the same set of principles. This is irrespective of whether the interviews were conducted with an individual or a group of individuals. Also, whether one approach is employed at a time or a complementarity of approaches is adopted, the same guidelines apply. This is to ensure and enable integrity, uniformity, and consistency in the data and entire process of data collection.

Guiding principles for data collection

To ensure and promote consistency, richness, and originality of data, the process of data collection needs to be guided by certain principles. Based on the discussion presented above, five factors are considered critical enough to be the guiding principles in data collection. These are (1) research ethics; (2)

interview guidelines; (3) pilot interviews; (4) selecting participants; and (5) demographics of participants. As shown by arrows in Figure 3.2, the principles are interconnected. Also, the principles are interdependent, which means that the phases must be followed sequentially from P1 to P5. However, phases four and five can be executed in parallel.

Ethical guidelines

Ethical considerations in IS studies reflect significant moral concerns and consequences about practice and responsibilities of researchers in the environments that they intend to conduct their studies. Universities have a research code of conduct which researchers, including postgraduates, adhere to. In some institutions, compliance is mandatory and determines project approval. A compliance certificate is issued to researchers to legitimise data collection. A code of ethics is used to guide research endeavours and to protect respondents. In case studies, ethics work in both ways – the institution and organisations where the study is being conducted. Many organisations, particularly the large ones, have ethical guidelines in the form of policies. The policies guide employees' behaviour in interacting with external bodies. No surprise why some employees or participants in studies are sometimes limited in the information that they share during interview sessions. Babbie (2007) describes ethics as conforming to standards of conduct in accordance with professionalism or the environmental culture. This includes respondents' responsibilities, such as assurance and protection of confidentiality and the right to participate in a study, or not do so. Thus, some participants prefer anonymity, in which case their identities are not revealed, and pseudo names are used.

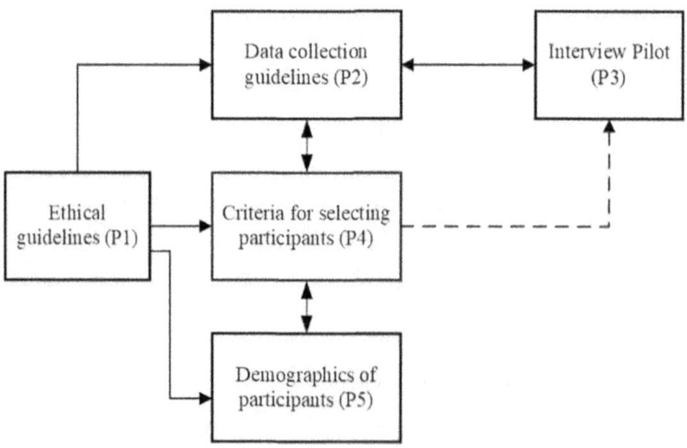

Figure 3.2 Guiding principles.
Source: Author.

Ethical standards and considerations include adherence to factors such as: (i) participant's consent and free will to enrol in an interview process; and (ii) a guarantee of confidentiality and anonymity in an effort to uphold research principles and to ensure that respondents' rights and values are maintained. In other words, information is treated with confidentiality and a commitment is made not to reveal respondents' identities. Information obtained during the course of the study is therefore strictly confidential to the researcher and the supervisor.

Data collection guidelines

In qualitative research, interviewers develop an interview guide to ensure depth, promote basic structure, and enable consistent focus on each of the interviews (Seidman, 2005). The interview guide consists of a list of questions that the researcher intends to pose to the interviewees. The interview guide is an integral tool in the process of collecting data in qualitative research. Without guidelines, consistency and uniformity become challenging or impossible to achieve. These are two crucial factors in the semi-structured interview technique.

The guidelines ensure that the researcher stays on track and consistently does so throughout the process. Other benefits of the guidelines are: (1) it is difficult to stay on track if the interviewee happens to have overwhelming evidence; and (2) where there is distraction, it becomes challenging to maintain consistency. McCracken (1988) argues that the use of an interview guide is essential for two main reasons: it ensures that the researcher covers all the terrain in the same order for each respondent and it establishes channels for the direction and scope of the discourse.

Pilot interview pilot

It is often suggested that researchers, particularly postgraduate students, should embrace the practice of a pilot interview. Fox-Wolfgramm (1997) explains why a pilot interview should serve as an initial guide for the entire interview process. This is more useful for those who are conducting an interview in academic research for the first time. The pilot interview helps to familiarise the interviewer with the interview guidelines and prepares them on how to probe responses. A pilot interview gives some advance warning in areas such as adherence to protocols, or appropriateness of methods or instruments. Also, piloting is essential because the researcher needs to be aware of the duration of the interviews and whether the directions and items are clear enough (McMillan & Schumacher, 2001).

Those that are selected to participate in a pilot interview should have similar interests or be in the same environment as those that will take part in the actual research. Consequently, the same criteria should be used to select those to partake in the pilot and those to participate in the actual interview sessions. Some of the objectives of the pilot interviews are to test the reliability and efficiency of the questions. Therefore, the pilot interview should assist the researcher to determine if there were errors, limitations, or other weaknesses in the interview questions. This allows the researcher to make necessary revisions to the questions and any other process that is involved. Also, the pilot interview can help the researcher gain more confidence in conducting the actual interview, as well as understand the environment better. These are contributing factors in gathering rich data.

Criteria for selecting participants

The types of participants that are selected in a study influence the integrity and richness of data. This makes criteria critical in the selection of participants. Therefore, random selection of participants is guesswork about participants' characteristics and should be avoided by all means. Otherwise, the interviewer will soon realise that some participants are indeed new to or less knowledgeable about the phenomenon being studied. The realisation reflects when participants begin to give short answers to questions or admit lack of knowledge in the subject. The criteria should include and highly consider factors such as experience (length of service), area of specialisation, and level in the organisational structure (or society). Depending on the type of study and its objectives, gender can also be part of the criteria for selecting participants to ensure a balanced view in a study. Also, there should be balance in other areas, such as senior and junior, according to organisational structure, in the selection of participants.

a Length of service – The rationale behind years of service as a criterion is to ensure that participants selected have a good or enough understanding of the environment. In the case of an organisation, this is to affirm that the participants are knowledgeable about the activities, such as strategies, structures, operations, policies, and procedures, and, thus, are able to give an informed opinion, and thereby provide valuable data.

b Area of specialisation – This is to ensure comprehensive and expanded views and options from the participants' perspectives. These are often or are expected to be based on their specialisation in the subject that is being studied.

c Departmental – The participants should be selected from different units or divisions of the organisation that is used as a case in the study. This includes participants' positions (roles and responsibilities) at the time of the study. This is to create balance in the selection of participants.

Demographics of participants

In IS studies, demographic information often focuses on geographical area, age, levels in the organisational structure, business unit, skill set, human race, gender, and education level. The demographics of the participants in a study is important, as it provides a vital explanation for the source of data. It promotes data richness and, most importantly, it instils confidence and originality in the data and the study in general. The demographics should also reflect balance among the selected participants.

Implication of practice

Based on the discussion as presented above, some factors were found to have implications for many postgraduate candidates in the course of practising the use of the semi-structured interview technique in IS research. The factors are of '-ship' nature. In this chapter, I define '-ship' as an indicative state or condition of peculiar purpose and context, in affinity. As shown in Figure 3.3, there are six main factors of implications, which are identified from the third section of this chapter: (i) the choice and selection of tools and their appropriateness for data collection in IS research; (ii) spoken language that is suitable for both the interviewer and the interviewees; (iii) 'situationship', a state of affairs which consists of inevitable or uncontrollable circumstances; (iv) 'probationship', a formal close examination or search into empirical evidence; (v) 'cultureship', a state of orientation and affiliation to traditional guidance; and (vi) 'locationship', a circumstantial place of physical presence or meeting.

Thus, the six factors are the main influences in the collection of qualitative data, using the semi-structured technique to answer the research question. Each of the six factors is explained below. It should be read along with Figure 3.3 to gain a better understanding.

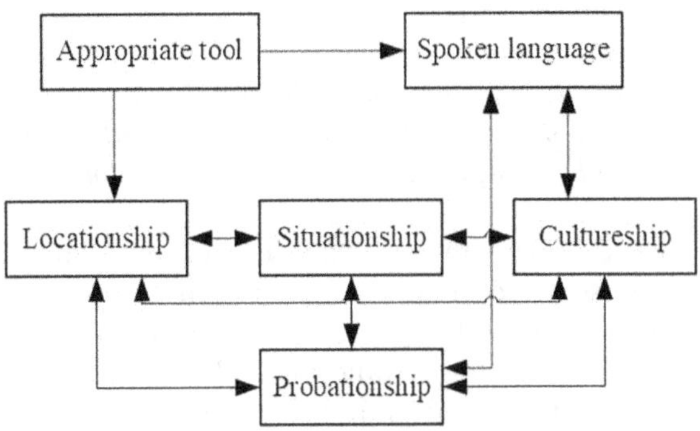

Figure 3.3 Challenging factors.
Source: Author.

Appropriate tools

The selection of appropriate tools for data collection is crucial in IS studies. What is even more critical is how well the tools are put to use mechanically or logically, or both. This is an area where many aspiring researchers or postgraduate students are often challenged. The challenge sometimes leads to confusion between cross-instruments in both qualitative and quantitative methods. So much work has been done in both research methods, yet the challenge persists. Some of the challenges are realised when a researcher selects a quantitative instrument and in the middle of an interview seeks to probe or get a clarification (explanation).

The selection and use of tools for data collection entail a process which requires that the aim and objectives of the study must be well understood, analysed, and assessed against available tools. Based on the tools, the researchers prepare themselves on the use and choice of spoken language, which is different from written language. They also select a location that is suitable to the interviewee. For example, some people are shy to express themselves in a loud manner in public or in an area where others can hear them. In the same vein, some people would prefer a location where they can freely speak in their local dialect, making them more expressive of their thoughts, views, and experiences.

Spoken language

Spoken language inextricably provides insight and meaning of things and actions on face value. In the deeper sense of things, it is more about processing of words and tenses from language rather than the output of it. It is therefore complex as it underlies symbolic systems that humans use to reproduce and create meaning to make sense of the social world. Postgraduate students who collect data in communities in which they are not fluent or do not understand the spoken language of the areas are highly likely to encounter challenges. This is because the student *takes for granted the medium of communication in the areas where data is to be collected. In this case, half of the people who could provide useful information for the study would not understand the questions or what is expected of them. Obviously, the community members or participants would like to converse in their local dialect.*

Also, there is a certain manner of speaking expected from visitors, including researchers, in some communities, which cannot be taken for granted. These are factors that are often guided by cultural belief and orientation. Thus, researchers need to understand and relate the environment within which they carry out their studies, in achieving the objectives of the phenomena under study. This will enable them to have insight on how they can associate meanings to words and things, towards gathering quality data.

Cultureship

A range of activities and ideas that are commonly shared as tradition by groups of people, transmitted and reinforced over a period of time. The establishment is carried out through spokespersons and mobilisation to enrol other interested persons, by virtue of the connectivity and relationship that they share. This is an established event over a period and space designed to gain stability of cultureship. Culture underpins the behaviour and character of many rural dwellers in many communities, particularly in the African continent. For example, when a younger person exchanges pleasantries with an elder through a handshake, the latter takes offence. A student once shared her experience with me: 'I used one hand to shake an elder's hand as opposed to using my both hands; as a result, I was ordered to leave the community without achieving my objective'. In another incident, a student explained how he was warned and asked to stop looking into the eyes of the interviewees who were married women.

It is therefore critical for researchers to understand the culture of an environment which they select to carry out their studies. Culture influences and impacts how certain events and actions are communicated and received in different situations. This must be understood by the researcher, as it is valuable to the richness of qualitative data.

Situationship

A state of situationship is a combination of circumstances that includes a place of event and relationship with actors (human or non-human). This is based on and influences the interaction between the actors, interviewer and interviewees, within the context of a study. Also, situationship is influenced by the comfort and interest of the actors as they connect or disconnect with an environment. Thus, some interviewees often have a preference about the location where they would like to be interviewed. This is as a result of certain or unique circumstances and situations at the time. Some participants would like to be interviewed at any place except their office premises. This is mainly because the interviewees have the intention to share personal views about the politics of the IT environment, which they perhaps consider confidential. In such circumstances, the interviewees are aware that superiors are vehemently opposed to the intended views, and fear victimisation. In contrast, some interviewees would like to be interviewed in an environment where they can demonstrate or show evidence of the views and insights that they are about to share.

The interest of interviewees varies, including inquisitiveness to know more about the study and an opportunity to gain some learning and knowledge. The interest could be based on voluntary or contractual willingness to contribute to the study. Situationship can be influenced by the location where the study is being conducted. Situationship also influences and has an impact on how the interviewer probes interviewees' responses.

Probationship

Probationship is an act to prove or test a scenario through subsequent interrogation and instant enquiry into a subject matter. Many researchers are sometimes challenged by the ability to instantly articulate follow-up questions, as they seek clarification and further exploration into participants' responses. This is often taken for granted and it is usually not practised before an interview process begins. *Instant articulation of follow-up questions to responses might be considered as a type of special skill. Despite theorising the challenge, many aspiring researchers or postgraduates struggle with this aspect during data collection, using the semi-structured interview technique. Some interviewers remedially prepare follow-up questions to interviewees' responses and request for a follow-up interview.*

One of the main benefits of the semi-structured interview is that it allows instant probing of interviewees' responses. However, many interviewers struggle with instant formulation and articulation of questions from the interviewees' responses. This often has an impact on the quality of data, which is mostly noticed during analysis.

Locationship

A venue that is suitable for or capable of hosting an interactive meeting to share information reflectively is referred to as locationship. Thus, a location where an interview is conducted is critical, but it is sometimes taken for granted by many interviewers and interviewees. Thus, some interviewers and interviewees are not indifferent about the location of the interview. However, some interviewees can be sentimentally specific about the location where they grant an interview. For example, a patient would like to be interviewed in a hospital room where they stayed for weeks in order to reflect on the real-life situation with medical personnel at the time. For some interviewees, it is an opportunity to reflect on reality within context rather than what could lead to fallacy in the process. Also, some interviewees see the interview as a prospect to contribute their unique experiences to academic studies. An interview, especially when it requires instant probing, at the right 'stage' can sometimes be almost therapeutic, and invites reflection, memories, and conclusions.

These factors, as discussed above, pose challenges in the collection of qualitative data, using the semi-structured interview technique in IS studies. What is even more challenging, and derailing, is the fact that some factors can hardly be separated from each other, as shown in Figure 3.3. One challenging factor can reproduce itself, thereby potentially manifesting and creating another challenge. For example, it is difficult to disconnect spoken language from the environment it takes place.

Conclusion

Although many works have been done in this area of qualitative methods – semi-structured interview technique – there will continue to be learning from new things that reproduce themselves in the process, at one point or the other. From this viewpoint, this chapter serves as education material for both learners and educators to consult. The chapter contributes to both the society and academia, particularly to aspiring researchers in the area of qualitative research. On the societal front, those who get to participate in a study will now get to understand why researchers ask certain questions in the manner that they do. On the academic front, the researchers will learn the significance of certain factors, such as spoken language, cultural affiliation, and certain situations in the process of using the semi-structured interview technique in an environment that they do not understand well enough.

References

Avison, D., & Malaurent, J. (2014). Is theory king? Questioning the theory fetish in information systems. *Journal of Information Technology, 29*(4), 327–336.

Babbie, E. (2007). *The practice of social research.* Belmont: Thomson.

Baškarada, S. (2014). Qualitative case studies guidelines. *The Qualitative Report, 19*(40), 1–25.

Bélanger, F., & Crossler, R. E. (2011). Privacy in the digital age: A review of information privacy research in information systems. *MIS Quarterly, 35*(4), 1017–1042.

Bryman, A. (2012). *Social research methods* (4th ed.). New York: Oxford University Press.

Choy, L. T. (2014). The strengths and weaknesses of research methodology: Comparison and complimentary between qualitative and quantitative approaches. *IOSR Journal of Humanities and Social Science, 19*(4), 99–104.

Cleary, M., Horsfall, J., & Hayter, M. (2014). Data collection and sampling in qualitative research: Does size matter? *Journal of Advanced Nursing, 70*(3), 473–475.

Creswell, J. W., & Poth, C. N. (2017). *Qualitative inquiry and research design: Choosing among five approaches.* London: Sage Publications.

Denzin, N. K., & Lincoln, Y. S. (2000). *Handbook of qualitative research.* London: Sage Publications.

Fox-Wolfgramm, S. J. (1997). Towards developing a methodology for doing qualitative research: The dynamic-comparative case study method. *Scandinavian Journal of Management, 13*(4), 439–456.

Goldkuhl, G. (2012). Pragmatism vs Interpretivism in qualitative information systems research. *European Journal of Information Systems, 21*(1), 135–146.

Hussain, Z., Wallace, J., & Cornelius, N. E. (2007). The use and impact of human resource information systems on human resource management professionals. *Information & Management, 44*(1), 74–89.

Katz, J. (2015). A theory of qualitative methodology: The social system of analytic fieldwork. *Methods: African Review of Social Sciences Methodology, 1*(1–2), 131–146.

Klein, H. K., & Myers, M. D. (1999). A set of principles for conducting and evaluating interpretive field studies in information systems. *MIS Quarterly, 23*(1), 67–93.

McCracken, G. (1988). *The long interview.* Newbury Park, CA: Sage Publications.

McMillan, J. H., & Schumacher, S. (2001). *Research in education. A conceptual introduction* (5th ed.). New York: Longman.

Myers, M. D. (2009). *Qualitative research in business & management.* London: Sage Publications.

Myers, M. D. (2013). *Qualitative research in business and management* (2nd ed.). London, Britain: Sage.

Myers, M. D., & Newman, M. (2007). The qualitative interview in IS research: Examining the craft. *Information and Organization*, *17*(1), 2–26.

Polak, L., & Green, J. (2016). Using joint interviews to add analytic value. *Qualitative Health Research*, *26*(12), 1638–1648.

Schultze, U., & Avital, M. (2011). Designing interviews to generate rich data for information systems research. *Information and Organization*, *21*(1), 1–16.

Seidman, I.E., (2005). *Interviewing as qualitative research.* New York: Teachers College Press.

Sekaran, U. (2003). *Research methods for business* (4th ed.). New York: John Wiley & Sons, Inc.

Silverman, D. (2013). *Doing qualitative research: A practical handbook* (4th ed.). London, Great Britain: Sage.

Stokes, P., & Urquhart, C. (2013). Qualitative interpretative categorisation for efficient data analysis in a mixed methods information behaviour study. *Information Research*, *18*(1), 1–17.

Tsang, E. W. (2014). Case studies and generalization in information systems research: A critical realist perspective. *The Journal of Strategic Information Systems*, *23*(2), 174–186.

Walsham, G. (2006). Doing interpretive research. *European Journal of Information Systems*, *15*(3), 320–330.

4 Activity theory

Table of contents

Introduction

Activity theory (AT) is a sociotechnical theory which originates from the field of psychology (Yamgata-Lynch, 2010) but has increasingly been applied in information systems (IS) research over the last three decades (Karanasios, Allen & Finnegan, 2015; Zins, 2017). This could be attributed to the theory's primary focus on social activities, which are indeed paramount in the development, implementation, and use of information technology (IT) solutions. McMichael (1999) explains that AT is increasingly popular as it focuses on the activities of people, which involves interaction and relationship. From an IS viewpoint, the theory has been applied in many studies, from different perspectives, to gain, describe, and explain the logics behind humans' activities in systems and the social world (Engeström, Lompscher & Rückriem, 2016; Shaanika & Iyamu, 2015). AT is argued to ease the explanation of humans' activities primarily because of the comprehensiveness of its components, which are tools, object, division of labour, community, rules, and subject (Engeström, 1990). The components focus on distinct areas, but they are interconnected in a phenomenon under study.

In AT, tools are constantly evolving, influenced by innovations and advancements. According to O'Leary (2010), tools shape the way in which people interact and connect with one another. An object is defined as the element or problem to which an action is directed or shaped. Nardi (1996)

DOI: 10.4324/9781003184119-4

explains how an Object is a component that guides how an entity is moulded and transformed into outcomes. A subject can be referred to as an individual or group of actors performing an activity. In Uden (2007), a Subject is referred to as both technical and non-technical entities that are engaged in an activity. The Rules are a set of policies and regulations that govern actors in their activities over a period of time, and within context. According to Burnard and Younker (2008: 62), 'rules historically govern or place constraint on actions within activity. Community is a group of people working towards achieving a common goal'. Blin and Munro (2008: 476) explain that community is composed of participants sharing the same object or motive. In AT, specialisation is defined as a form of Division of Labour, a component that is concerned with how tasks are divided among community members in carrying out an activity. Due to the power that actors possess, they can enable or constrain systems and technologies' activities based on their conscious or unconscious actions.

In IS research, AT is applied to focus on studying human activities within social systems. From this perspective, Engeström et al. (1999) explained that human activity is composed of three levels: activity, action, and operation. An activity is decomposed into actions and each action is decomposed into operations, which can include the development, implementation, and management of IT solutions (Moawad, Liu & El-Helly, 2003). According to Nardi (1996), each activity is performed through either individuals or corporates, including chains of actions that are related to one another towards a common goal. In Golsorkhi et al. (2010), it is argued that the theory 'conceptualises the on-going construction of activity as a product of activity systems comprising the actor, the community within which that actor interacts, and those symbolic and material tools that mediate between actors, their community and their pursuit of activity' (p. 127).

In AT, living beings are referred to as actors (Engeström et al., 2016), which is different from the actor-network theory's (ANT) definition of actor, where actors are both human and non-human (such as technical entities). Actors are entities that have the ability to make a difference in ANT (Iyamu, 2015). In the context of AT, actors have significant roles and contributions to activities within a social system or community. From an ANT perspective, Thapa (2011) explains that each actor has the ability to act and also alter, which means that both human and non-human actors are inseparable. Actors make use of tools to perform various activities towards achieving goals and objectives. Tools can be physical entities, such as a hammer, or non-physical e.g., psychological, which includes language, culture, or ways of thinking (Uden, 2007). The inseparable nature of both technical and non-technical actors is enacted by process and people.

The development, implementation, and management of IT solutions are carried out through processes that are formulated by people (Iyamu, 2013). Development and implementation entail processes in activities through which IT solutions are introduced into an environment. The processes can also be used to reject IT activities (Allen, Karanasios & Slavova, 2011). The processes and activities that are involved in the development, implementation, and management of IT solutions are (1) subjectively defined; and (2) subjectively understood by the different actors. A case in point is Carvalho et al. (2015), where AT was used to explain the subjective roles of actors in the development and implementation of IT solutions. Hence, steps and guidelines are critical to the success or failure of IT solutions where AT is applied.

In recent years, AT has been applied as a lens for data analysis and interpretation, in the field of IS, it is predominantly in qualitative studies (Karanasios et al., 2015). However, there is no formal approach or scholarly guide that is followed in applying the theory in IS studies. Primarily, this is rationale and focus of this chapter, based on which the question is posed: What approach can be used to guide applying AT for analysis or interpretation of qualitative data in IS studies? It requires a more in-depth enquiry to find answers to this question. The qualitative research method allows the use of interrogative questions, which begin with interrogatives 'who, what or how', to assess and evaluate qualitative data through analysis process. This is intended to ease analysis or interpretation of data in complex situations. Dane (2010: 12) proposes six components (who? what? where? when? how?, and why?) as an approach for the assessment and evaluation of phenomena in a research.

The remainder of this chapter is divided into five main sections. The first and second sections present discussions about AT; and the third, AT and IS, respectively. AT as a lens is covered in the fourth section. A guide on how to apply AT for analysis or interpretation of qualitative data in IS research follows. Finally, a conclusion is drawn.

AT overview

AT focuses on studying and understanding the context of human activity within their socio-environment (Karanasios et al., 2015). The focus is observed from the constituent of AT, which are a subject and an object, mediated by a tool (Engeström et al., 1999) including community, rules, and division of labour (Shaanika & Iyamu, 2015). Together, they denote the social context where activities are carried out (Uden, 2007).

From the perspective of AT, Kaptelinin and Nardi (1997) argue that actions by actors are often conscious. Within the same concept, Nardi (1996) emphasised that consciousness is a basic principle of AT, as the theory seeks to understand the unity of consciousness and activity. Consciousness is described by Kaptelinin and Nardi (2006: 9) 'as the enactment of our capacity for attention, intention, memory, learning, reasoning, speech, reflection and imagination'. Along the same viewpoint, Moawad et al. (2003) argued that before actions are performed, they are planned consciously.

The concept of AT holds that the constituents of activity are not fixed but can dynamically change as conditions change (Kaptelinin & Nardi, 2006).

Thus, McMichael (1999) explains that the subject of an activity is always active and never passive and evolves over a period of time. According to Moawad et al. (2003), activities evolve in a non-linear way within a social system. In the same context, Engeström et al. (1999: 4) describes activities as 'open systems' and explains that when an activity is undertaken within a system, it produces a new element from the outside (e.g., new technology or innovation), which often leads to an aggravated secondary contradiction where some old element (e.g., rules or the division of labour) collides with the new one. In Nardi's (1996) view, such contradictions not only generate disturbance and conflict, but also innovate attempts to change the activity.

As shown in Figure 4.1 and discussed below, AT depicts the activity context as a network of different elements (subject, object, tools, division of labour, community, and rules) that influence each other, in a triangle model (Engeström et al., 1999).

Tools can be any physical or psychological entity, including language, culture, or ways of thinking (Uden, 2007). In a complementary view, Moawad et al. (2003) explain that tools can be a sign, procedure, machine, method, or law. According to Holt and Morris (1993), tools mediate the subject activity towards the object. Tools are used by subjects when performing activities.

A subject consists of technical and non-technical entities that are engaged in activity operations (Shanika & Iyamu, 2015). McMichael (1999) defines subject as the agent that acts upon the object of the activity. A subject can be referred to as an individual or group of actors who perform an activity.

The object can be a material thing, less tangible, or totally intangible (e.g., idea) as long as it can be shared by those involved in an activity (Karanasios et al., 2015). According to Nardi (1996), activity is directed at an object which motivates an operation. Engeström et al. (1999) explains how activities become object-oriented in a social system. However, it has been emphasised

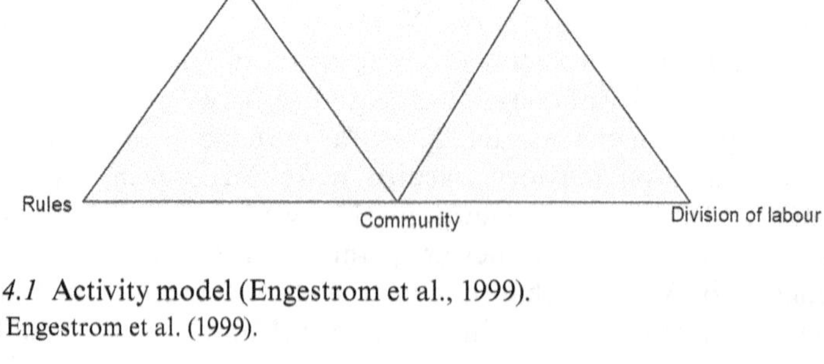

Figure 4.1 Activity model (Engestrom et al., 1999).
Source: Engestrom et al. (1999).

not to confuse object-oriented with the same term used in computer programming (Er, Kay & Lawrence, 2010; Holt & Morris, 1993). Engeström et al. (1999) noted that an object provides activity with a specific direction in a social system.

Community is an interdependent aggregate of individuals who share a set of social meanings (Allen et al., 2011). Yamgata-Lynch (2010) describes community as a social group within which the subject is identified while participating in an activity. Uden (2007) argues that it is of importance to know and take cognisance of the fact that the individuals that are involved in a particular activity are simultaneously members of activities in other communities that have different objects, tools, and social relations.

Activities are carried out in communities in accordance with rules (Karanasios et al., 2015). Yamgata-Lynch (2010) posits that in an analysis of activities, different kinds of rules, laws, norms, and cultural practices are also taken into consideration about the community where an activity occurs. Rules govern subjects as they carry out their activities and interactions with other components of the community (Carvalho et al., 2015). Thus, on the one hand, rules are necessary for maintaining order in communities. Without rules it would be difficult to achieve defined objectives in a community. On the other hand, rules can, in varying degrees, constrain or liberate an activity (Yamgata-Lynch, 2010).

Division of labour represents the allocation of specific tasks in accordance with areas of specialisation of individuals (Holt & Morris, 1993). Division of labour is concerned with allocation of tasks among community members and how the allocated tasks manifest into source of power and status in an environment (Shaanika & Iyamu, 2015). Division of labour is crucial, as this enables the sharing of activities among community members, in which each subject assumes responsibility for an activity.

An outcome is the result from an activity. Each component of AT significantly contributes and has an impact on the outcome of an activity in a social system. The impact can be positive or negative, depending on the situation or the phenomenon being studied or the environment where the activity is carried out (Allen et al., 2011).

AT and IS

For many years, AT has been used in IS research to study human activities in their natural settings. Crawford and Hasan (2006) argue that one of the characteristics of AT is that it facilitates interaction between human and technical elements in an activity. Thus, an activity is considered to consist of an occurrence that constitutes essential qualities and characteristics of human actions, interaction, relationship, and practices. In addition, an activity is also seen as a form of event that focuses on an object within which actions are carried out in accordance with an actor's objectives and responsibilities.

AT is used in IS research increasingly to provide an understanding of how humans interact with other actors in their environments when developing, implementing, and using IT solutions (Iyamu & Shaanika, 2019). Increasingly, applying AT in IS research or practice often focuses on individuals or groups, to reveal the determinants and consequences of their actions in the development or implementation of IT solutions. Dennehy and Conboy (2017) suggest that the use of AT helps to identify contradictions that cause problems such as errors or a breakdown in communication. Therefore, it can be argued that AT enables a deconstruction of an activity with its original complex nature during analysis. This is purposely to simplify and gain a better understanding of the nature of works, roles, and responsibilities of actors that are involved in an activity (Nardi, 1996). Also, AT helps to define and distinguish between components, processes, and their relationships within an activity (Er et al., 2010). Mursu et al. (2007) suggest that the basic unit of analysis in the use of AT is an activity that connects individuals' actions to context. Thus, the theory is able to assist in revealing flow as well as contradictions between components (Dennehy & Conboy, 2017).

Therefore, AT is somehow considered a framework because it is used to enhance design practices in Computer Supported Collaborative Work (CSCW) and other related fields, such as Human-Computer Interaction (HCI). This affirms Nardi (1996), where AT is described as a theoretical framework for explaining the structure, development, as well as social context of human activities within an environment. AT is a theory-based context analysis technique that anchors the ethnographer's descriptive work, calling their attention to the individual and social elements of human activities.

AT is often applied in IS studies mainly because the field is part of social environments designed and managed, using tools, by people. According to Sekgweleo et al. (2017), the theory has been applied over 3 million times in IS research. AT is usually employed in IS research to examine the activities of people as they interact with each other in an attempt to achieve desired outcomes (McMichael, 1999). Most importantly, AT is often applied as a lens in IS studies to guide the analysis of data, usually from a qualitative perspective (Karanasios et al., 2015). The use of AT for analysis is normally to develop theories in the form of a model or framework. The theory acts as a third eye used to examine and understand how responses or events manifest from questions prefixed by what, how, and why non-technical and technical factors impact and influence the activities of IT solutions in an environment.

From an AT perspective, activities are categorised into units for analysis purposes. The units of analysis are neither individuals nor the environment, but a relationship between the two entities (Nardi, 1996). As argued by McMichael (1999), unless the whole activity is the unit of analysis, the analysis is incomplete. As it is in IT solutions, no action or activity is complete by a single entity or actor. For example, technology needs humans and vice versa, before any action can be completed. According to Kaptelinin and Nardi (1997), activities are realised as multiple individual and corporate actions which cannot be understood unless they are related to the overall objective and motive of an activity in a social system. Every aspect that has an impact on an activity is considered in the analysis of the activity. This is done in a subjective manner since an individual or group of analysts' senses making is required in the phenomenon being studied.

The interpretivist approach allows the researcher to analyse the interviewees' experiences and subjective reasoning (Rowley et al., 2011). The core idea of interpretivism is to fathom the subjective meanings which persons associate with things (Goldkuhl, 2012). To attain this, the researcher uses his or her skills as a social being to try and comprehend how others understand or relate to the world around them (O'Donoghue, 2006). The interpretivist approach is often used in IS studies purposely to gain an understanding of how employees interact in the selection, implementation, governance, management, and use of IT solutions in an organisation.

Also, IS researchers employ AT in their studies primarily because the development, implementation, and management of systems and technologies (or IT solutions) are regarded as social-technical activities (Allen et al., 2011). The use of AT to underpin the development or implementation of IT solutions in an organisation focuses on three main areas: (i) how actors are involved; (ii) how actors' involvement is governed by an organisation's cultures, values, and norms, based on requirements; and (iii) how division of labour is upheld through the allocation of roles and responsibilities. The governance is mainly to ensure and enforce individuals' commitment, focus, and orderliness. In addition, governance is employed from critical aspects that are needed to achieve organisational goals and objectives. However, rules can come in varying degrees, and can constrain or liberate the activity (Yamgata-Lynch, 2010).

AT as a lens in IS research

AT is usually employed as a lens to analyse and interpret data in IS research. According to Walsham (2006), a theory can be used as an initial guide to a study design and as part of an iterative process of data collection and analysis. From a significance point of view, Sekgweleo et al. (2017) explain that the theory is often selected to underpin IS research primarily because it provides a conceptual framework for understanding and analysing human activity from the perspectives of both humans and technical factors within computing environments. In AT, humans and technical factors depend on each other for the fulfilment of goals and objectives within a community in an organisation (Shaanika & Iyamu, 2015). An attempt to separate humans from technical factors in executing IT solutions creates a vacuum. For example, without a technical factor, such as a computer, it is difficult or impossible for humans to process thousands of customers' claims.

Some of the technical (IT solutions) factors are financial systems, biometric systems, computer servers, databases, different types of networks, and personal computers (PCs). The technical factors could also be referred to as tools, in that they are used to achieve objectives. This implies that an IT solution is used by humans (actors) in the same environment to achieve an objective. The actors are stakeholders, such as vendors, and consultants from different organisations, and support staff, managers (such as chief executive officers, and business directors) and IT specialists, which include systems analysts, software developers, directors, and network administrators,.

Even though AT has been employed in over 3 million of IS studies for many years, it continues to pose challenges to many researchers, particularly aspiring researchers and postgraduate students. Consequently, the challenges prompt a helpful guide on why and how to apply the theory. This chapter focuses on these challenges and it is intended to create an awareness for postgraduate students and aspiring researchers, thereby continue to promote interests in the theory for IS studies' purposes. Overcoming the challenges is not a one-off activity but rather an iterative process. Primarily, this is so because during analysis, the researcher analyses the data back and forth, resulting in new terms being discovered. Figure 4.2 is presented as a possible solution to the challenges that are often encountered in applying AT in IS research, articulated as follows: (1) lack of better understanding of the theory; (2) misperceptions during data analysis; and (3) lack of an

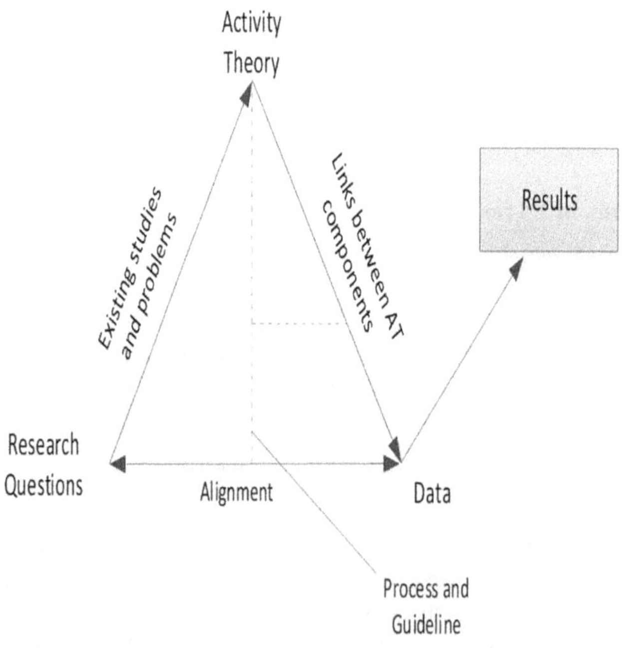

Figure 4.2 AT for data analysis.
Source: Author.

understanding of the linkage between the components of the theory. This should be read with reference to Figure 4.2 to help readers have better comprehension of the intention.

Have a good understanding of the theory

In order to make use of AT as a lens in IS studies, the theory should be well understood. Better understanding refers to the individual's knowledge of the theory and its tenets as shown in Figure 4.1. An understanding refers to the type of knowledge that individuals gain from an activity, which can be of a theoretical or practical nature. More often than not, many researchers have the theoretical understanding of AT, but lack the knowledge that is required to make use of the theory for practical purposes. The ability to align with two key factors to ensure appropriateness of the theory in an IS study is always required: (i) research questions; and (ii) gathering of qualitative data. The use of AT does shape and have a significant impact on the results from a study.

This challenge can be attributed to the fact that AT is not originally an IS subject; it has its roots in physiology, a field which many students and researchers of IS are neither trained to understand nor conversant with. However, due to the sociotechnical aspect of IS, the continued need to enquire and gain a fresh perspective of why things happen in the way that they do in IS environments makes it essentially viable and increasingly important to employ AT for analysis.

Self-misperception in the use of AT

In many situations, researchers believe they have a good understanding of AT. This remains the self-certainty until such time the theory is being applied to guide the analysis of data. At that point, some researchers (or students) realise how little they do know about the theory in applying it as a lens. This type of self-misconception sometimes creates confusion, causing some aspiring researchers to think that perhaps they collected insufficient or irrelevant data. In such scenarios, some researchers do have the hunger and zeal to revisit their data collection process and, if possible, do more. The major implication of such circumstance is that the researcher is likely to be influenced and guided by the theory, as opposed to the research questions and objectives.

Should data collection be based on or totally influenced and guided by AT, it changes the scope and objectives of a study and creates gaps in the outcome. Sekgweleo et al. (2017) comprehensively explain how the selection of the theory should be guided by the research questions and must be linked to the study objectives. Therefore, the data collected should be suitable for AT in that the selection of the theory is linked to the research questions. The main cause for the misconceptions during data analysis is likely to be attributed to lack of good understanding of how individual components of the theory are used and interlinked in the process of applying AT.

Understanding the interconnection between the components of AT

The use of AT as a lens to guide the analysis and interpretation of qualitative data in IS research follows a process in which each tenet of the theory is linked with sets of data. The link between the data and AT enables the researcher to examine and gain an understanding of the study, fundamentally, these are: (i) what tools (such as software, network protocols, and hardware) are employed in carrying out specific activity; (ii) why certain types of objects are involved in an activity; (iii) how and why rules are formulated and applied; (iv) how actors' actions evolve and manifest in various activities to shape outcomes; (v) how and why various communities (such as groups of employees and teams of specialists) are formed in an environment; and (vi) the reasons why tasks are allocated and carried out in the manner that they are in an environment. Through these links, which are analysed following the interpretivist approach, results are produced, and findings are reached which detail how and why activities reproduce themselves in a social system.

The six components are not independent of one another. As a result, it is near impossible to analyse data in an activity without involving all components and not create a vacuum that will prohibit achieving the objectives of the study. It does require a holistic coverage of the components and their interconnectivity within the context of the phenomenon being studied to further induce the knowledge about AT. This is an important aspect in that the components cannot be applied in isolation, but rather in an interconnected approach. Such knowledge enables and allows a researcher to establish and recognise how each component influences an activity. The links between the components are the catalysts that are needed to successfully carry out data analysis in the use of AT. By implication, the approach enables a researcher to gain a better understanding of how the output of one component can be used as an input for the preceding components. This serves as guidelines for applying the theory and promotes ease of data analysis in IS studies.

A guide on using AT

This section presents a guide on how to apply AT as a lens for data analysis in IS studies. As shown in Figure 4.3, the guide is a three-dimensional phase (P1 ... P3) approach, which are P1 – selecting AT for an IS study; P2 – elements of the analysis; and P3 – linking the elements of the analysis with AT components. The approach should be applied in sequential order to achieve the objectives of the study.

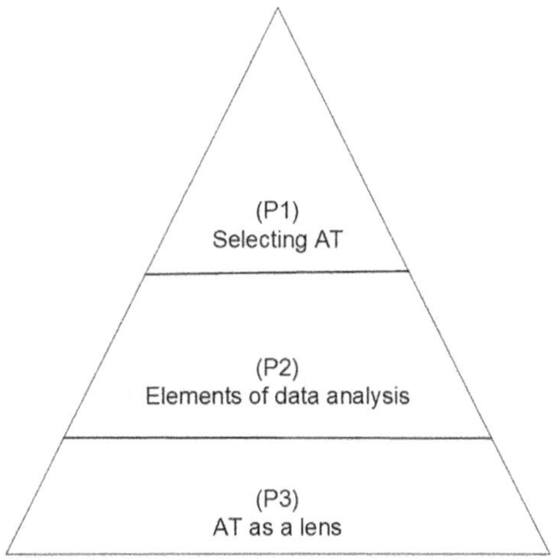

Figure 4.3 A guide for AT.
Source: Author.

Selecting AT for IS studies

The objectives of a research are the primary drivers that guide the selection of AT for the purposes of data analysis in IS research. As revealed and discussed in sections, "AT and IS" and "AT as a lens in IS research", respectively, four main factors form the basis of the activities that are undertaken in the development, implementation, or management of IT solutions in organisations. The factors are: (i) interaction and relationship; (ii) technical know-how; (iii) networks of people; and (iv) governance and structure. Hence, many IS studies are either interrelated or interconnected to any or a combination of the factors, primarily from one technological perspective to another.

i ***Interaction and relationship:*** interaction and relationship between subjects and objects are inseparable in IS research, whether in development, implementation, or management. Interaction happens because relationship exists, and vice versa. On these bases, the researcher assesses the research objectives, to establish the implications of interaction and relationship between human and non-human, human-to-human, human-to-technology, or technology-to-technology concerning the phenomenon that is being studied.
ii ***Technical know-how:*** defines the nature of tasks, including the capacity and capability that are required to fulfil activities within an environment. This is used to further explain how tasks are defined, allocated, and managed towards achieving the objectives of each activity in the

development, implementation, or management of IT solutions. Also, the technical know-how helps to identify and categorise an available skillset in the development, implementation, or management activities.

iii ***Network of people:*** distinguishes and categorises the different groups of people that exist within an environment. This helps to create boundaries within which various activities take place. The boundaries can be created through three different means – formally, informally, or practical consciousness, which Giddens (1984) describes as individuals' ability to communicate actions and motivations explicitly in terms of their own knowledge. Examples of network of people are teams of software developers, a race group, and a group of people with common interest, such as using a specific programming language for software development. Each network has the capability to make a difference in their environment, from development, implementation, or management perspectives.

iv ***Governance:*** this entails the enforcement of rules and regulations in the development and implementation of IT solutions. In many IS studies, rules and regulations deterministically influence the activities (e.g., support and maintenance) of IT solutions in an environment.

Many researchers in the field of IS who employ AT as a lens do so to gain knowledge of how and why humans interact with each other or with technology in their environments, specifically in the development, implementation, or management of IT solutions. The theory focuses on actors' interactions and relationships with each other in a social system (community), by using different materialistic tools. Therefore, AT enables the decomposition of a complex activity during analysis of data so that a detailed understanding of the nature of works and responsibilities of those involved in the activity can be gained (Nardi, 1996). Also, it helps to establish the kinds of components and processes that are incorporated within an activity and helps to define the relationships that exist between subjects (Er et al., 2010).

Within the factors that determine the selection of AT, questions are often asked to assess or evaluate how activities came to exist and manifest to enable and constrain the tasks of the IT solutions. In the formation of the questions, prefixes are proposed and used to elicit detailed constructivists' information. This is an approach that can be employed to analytically examine activities in IS research.

Elements of data analysis

Table 4.1 presents the elements which are fundamental to the analytic scheme in the use of AT as a lens for data analysis. As shown in Table 4.1, the use of the prefix guides applying AT as a lens, in an interrogative manner, to elicit detailed subjective retorts. Iyamu (2015) argues that these elements are fundamental to stimulating answers during the analysis of data. Dane

Table 4.1 Elements of the analytics

Elements	*Description*
What	Clarifies the object, subject, or tools that are being used or can be applied in an activity within the context and relevance in an environment.
Who	Establishes and identifies the actors that show interest, participate, or can enrol in an activity.
How	By implication, defines and interconnects the actors that are involved in an activity.
When	This defines the context or period within which an activity takes place.
Where	The location, boundary, and scope within which an activity takes place are identified and defined.
Why	Helps to induce the enabling and constraining factors in an activity.

(2010) systematically employs the elements 'who, what, when, how, where, and why' for assessment and evaluation of a system in qualitative research.

The elements, as shown and described in Table 4.1, are analytically used to elicit information about IT solutions in IS studies. This is purposely to gain insights about why things happen in certain manner during development, implementation, or management of IT solutions.

Components of AT

As a lens, each of the components of AT is usually intended to bring fresh perspectives in assessing or evaluating an activity that occurs in the development, implementation or management of IT solutions. This is often with one purpose – to stimulate or bring forth new knowledge about the phenomenon that is being studied. From an IS research perspective, the six components of AT are briefly described as follows:

i Tool – this helps to establish the tools (relating to IT solutions) that are deployed (developed or implemented) and used from the main points of: (1) where the tools are deployed and used; (2) how the tools are governed, supported, and managed; and (3) why the tools are deployed and used in the manner that they are in a computing environment.

ii Subject – this helps to examine and comprehend the subjects that exist within an environment. It includes how the subjects interconnect, interrelate, or interact with each other in the process of enabling and supporting goals and objectives of an organisation.

iii Object – through this component, the existence of objects is revealed or established. Also, it brings different viewpoints in discovering the ultimate role of objects in an environment. This relates to the deployment, use, and management of systems and technologies.

iv Rule – this component helps to assess how guidelines and regulations are promulgated and used to govern related tasks of IT solutions. The governance is through either negotiation or an obligatory passage point (OPP), in the selection, implementation, or management of systems and technologies within an environment. OPP is broadly defined as a situation that has to happen for actors to satisfy the interests that have been assigned to them by the focal actor (Iyamu, 2015).

v Community – focuses on the different networks of people that exist within an environment. This component is employed to gain insightful prepositions on how systems and technologies are enacted through existing networks, to address challenges or enhance solutions in the different areas of an organisation.

vi Division of labour – this component covers and is used to analyse how roles and responsibilities, including tasks allocation, are negotiated and carried out to achieve the organisation's goals and needs. This component can be used as a guide in evaluating the roles of the focal actor and the responsibilities of other actors in the deployment of IT solutions within an environment.

AT helps to assess and evaluate both internal and external variables or factors in the development, implementation, management, or use of IT solutions within a social system (an organisation). The activities that influence IT solutions are either internal or external bound. There is usually an emphasis that the internal activities cannot be analysed separately from external activities because they transform into each other (Nardi, 1996). In essence, the internal activities are the result of transforming external activities vis-à-vis Object and Subject as the actors that make use of the available tools.

As shown in Table 4.2, the elements of analysis (Table 4.1) are linked with AT components, to complete the three-phase approach that is proposed in this chapter. Each cell (junction between row and column) depicts AT as a lens that can be used to explicitly extract detailed and fresh perspectives from empirical data in IS research. Each cell is contextual and of specific relevance. The cells should therefore be read relative to the phenomenon that is being studied, where AT is applied. Thus, the tenses in cells may be rephrased or customised for suitability purposes.

Table 4.2 can be used to guide the analysis of both internal and external factors in the activities concerning IT solutions at strategic and operational levels. Moawad et al. (2003) explain how an activity can be decomposed into actions that are further decomposed into operations. Operations are the ways in which actions are carried out in an environment. Also, an activity is driven by a motive with actions, directed by objectives within context, which Table 4.2 tries to unpack sequentially and comprehensively. McMichael (1999) reported that human activities involve the use of instruments such as tools which mediate between subjects' actions. The utilisation of tools changes the structure of the activity and can result in new goals to

Table 4.2 AT Components for analysis

Component	*What*	*Who*	*How*	*Where*	*When*	*Why*
Tool	What tools exist in or for an activity?	Who makes use of the available tools?	How are the tools applied in an activity?	Where are the tools applied in an activity?	When are the tools applied in an activity?	Why are tools applied in an activity?
Object	What objects are available in executing an activity?	Who makes use of the available objects in executing?	How are the objects used in executing an activity?	Where are the objects used in executing an activity?	When exactly are the objects used in executing an activity?	Why are the objects employed in executing an activity?
Subject	What subjects are available in executing an activity?	Who are the subjects that exist in executing an activity?	How does a subject exist within or in executing an activity?	Where does a subject exist within or in executing an activity?	When does a subject exist within or in executing an activity?	Why are subjects involved in executing activities?
Rules	What rules guide the existence or execution of an activity?	Who makes the rules, including the rationale for an activity?	How are the rules formulated, promulgated or applied in an activity?	Where are the rules applied in an activity?	When are rules applied within an activity?	Why are the rules formulated, promulgated, or applied in an activity?
Community	What are the communities that exist in executing an activity?	Who are the communities that exist in executing an activity?	How does a community exist in executing an activity?	Where does a community exist in executing an activity?	When does a community exist in executing an activity?	Why are communities created or exist in executing an activity?
Division of labour	What tasks, labour, or division exist in executing an activity?	Who is involved in executing an activity?	How are tasks assigned to actors in executing an activity?	Where are the various tasks required in executing an activity?	When are the various tasks assigned in executing an activity?	Why are the various tasks assigned in the manner that they are in executing an activity?

be satisfied. Additionally, according to Nardi (1996), both the subject and the community are modified through the mediated activity.

Also, Table 4.2 can be used as a frame to guide the development of the AT curriculum for data analysis in a research methodology course. Through the curriculum, students need to understand that individuals do not only produce outcomes, such as research projects and theses, which are distributable and consumable knowledge, they, in the same process, reproduce themselves as members of the community (Roth, 2004). From research methodology perspective, this chapter should contribute to the development of the curriculum and the enhancement of student learning as well as professional practice (Leathwood & Phillips, 2000)

Conclusion

The use of AT as a lens to assess or evaluate activities in the development, implementation, management, or use of IT solutions in IS research is clearly critically important for education purposes, through research, teaching, and learning. As revealed in this chapter, this is primarily because the theory focuses on the interaction and relationship between humans and non-humans, which forms the underlining principles and factors of existence, including assessments and evaluations of IT solutions in organisations.

The three-phase approach proposed in this chapter is intended to guide both aspiring researchers and researchers who are new to AT in their selection and applying the theory in IS studies. The deterministic nature of research objectives that link the analytics elements with AT components is of peculiar contribution. In particular, Table 4.2 significantly eases analysis in that it helps to gain deeper elicitation of information from empirical evidence that would have been difficult to achieve. Table 4.4 can also be of significant value to both business and project managers in organisations to understand better why things happen in the way that they do in various activities.

Another contribution of this chapter is that it can ease how AT may be used as a lens to examine sociotechnical complexities, and to recognise and provide an understanding of the relationship that happens between human actors and technical factors. In addition, the chapter helps to unpack how the factors influence and impact each other in the activities of development, implementation, management, and use of IT solutions in organisations. Thus, the three-phase approach will guide researchers, particularly postgraduate students, in the use of AT in IS research.

References

Allen, D., Karanasios, S., & Slavova, M. (2011). Working with activity theory: Context, technology, and information behaviour. *Journal of the Association for Information Science and Technology*, *62*(4), 776–788.

Blin, F., & Munro, M. (2008). Why hasn't technology disrupted academics' teaching practices? Understanding resistance to change through the lens of activity theory. *Computers & Education*, *50*(2), 475–490.

Burnard, P., & Younker, B. A. (2008). Investigating children's musical interactions within the activities systems of group composing and arranging: An application of Engeström's Activity Theory. *International Journal of Educational Research*, *47*(1), 60–74.

Carvalho, M. B., Bellotti, F., Berta, R., De Gloria, A., Sedano, C. I., Hauge, J. B.,... & Rauterberg, M. (2015). An activity theory-based model for serious games analysis and conceptual design. *Computers & Education*, *87*, 166–181.

Crawford, L., & Hasan, H. 2006. Demonstrations of the activity theory framework for research in IS. *Australasian Journal of Information Systems*, *13*(2): 49–68.

Dane, F. C. (2010). *Evaluating research: methodology for people who need to read research*. London: Sage.

Dennehy, D., & Conboy, K. (2017). Going with the flow: An activity theory analysis of flow techniques in software development. *Journal of Systems and Software*, *133*, 160–173.

Engeström, Y. (1990). *Learning, working and imagining: Twelve studies in activity theory*. Helsinki: Orienta-konsultit.

Engeström, Y., Lompscher, J., & Rückriem, G. (Eds.). (2016). *Putting activity theory to work: Contributions from developmental work research* (Vol. 13). Berlin: Lehmanns Media.

Engeström, Y., Miettinen, R., & Punamäki, R.-L. (Eds.). (1999). *Perspectives on activity theory*. Cambridge: Cambridge University Press.

Er, M., Kay, R., & Lawrence, E. (2010). Information systems and activity theory: A case study of doctors and mobile knowledge work. In *Seventh International Conference on Information Technology: New Generations (ITNG), 12–14 April* (pp. 603–607). Las Vegas, NV, USA. IEEE.

Giddens, A. (1984). *The constitution of society: Outline of the theory of structuration*. Cambridge: Polity Press.

Goldkuhl, G. (2012). Pragmatism vs. interpretivism in qualitative information systems research. *European Journal of Information Systems*, *21*(2), 135–146.

Golsorkhi, D., Rouleau, L., Seidl, D., & Vaara, E. (Eds.). (2010). *Cambridge handbook of strategy as practice*. Cambridge: Cambridge University Press.

Holt, G., & Morris, A. (1993). Activity theory and the analysis of organizations. *Human Organization*, *52*(1), 97–109.

Iyamu, T. (2013). Underpinning theories: order-of-use in information systems research. *Journal of Systems and Information Technology*, *15*(3), 222–238.

Iyamu, T. (2015). *Application of underpinning theories in information systems*. Australia: Heidelberg Press.

Iyamu, T., & Shaanika, I. (2019). The use of activity theory to guide information systems research. *Education and Information Technologies*, *24*(1), 165–180.

Kaptelinin, V., & Nardi, B. (2006). *Acting with technology: Activity theory and interaction design*. Cambridge: Massachusetts Institute of Technology Press.

Kaptelinin, V., & Nardi, B. A. (1997). Activity theory: Basic concepts and applications. In *CHI'97 Extended Abstracts on Human Factors in Computing Systems* (pp. 158–159). ACM.

Karanasios, S., Allen, D., & Finnegan, P. (2015). Information systems journal special issue on: Activity theory in information systems research. *Information Systems Journal*, *25*(3), 309–313.

Leathwood, C., & Phillips, D. (2000). Developing curriculum evaluation research in higher education: Process, politics and practicalities. *Higher Education*, *40*(3), 313–330.

McMichael, H. (1999). An activity based perspective for information systems research. In *10th Australian Conference on Information Systems.*

Moawad, N., Liu, K., & El-Helly, M. (2003). Integrating activity theory and semiotics as knowledge elicitation technique. *14 ACIS International Conference on Software Engineering, Artificial Intelligence, Networking and Parallel/Distributed Computing.* IEEE Computer Society.

Mursu, A., Luukkonen, I., Toivanen, M., & Korpela, M. (2007). Activity theory in information systems research and practice: Theoretical underpinnings for an information systems development model. *Information Research*, *12*(3): 1–21.

Nardi, B. (Ed.). (1996). *Context and consciousness: Activity theory and human-computer interaction.* Cambridge: Massachusetts Institute of Technology.

O'Donoghue, T. (2006). *Planning your qualitative research project: An introduction to interpretivist research in education.* London: Routledge.

O'Leary, D. E. (2010). Enterprise ontologies: Review and an activity theory approach. *International Journal of Accounting Information Systems*, *11*(4), 336–352.

Roth, W. M. (2004). Introduction: "Activity theory and education: An introduction". *Mind, Culture, and Activity*, *11*(1), 1–8.

Rowley, J., Jones, R., Vassiliou, M., & Hanna, S. (2011). Using card-based games to enhance the value of semi-structured interviews. *International Journal of Market Research*, *54*(1), 93–110.

Sekgweleo, T., Makovhololo, P., & Iyamu, T. (2017). The connectedness in selecting socio-technical theory to underpin information systems studies. *Journal of Contemporary Management*, *14*(1), 1097–1117.

Shaanika, I., & Iyamu, T. (2015). Deployment of enterprise architecture in the Namibian government: The use of activity theory to examine the influencing factors. *The Electronic Journal of Information Systems in Developing Countries*, *71*(1), 1–21.

Thapa, D. (2011). The role of ICT actors and networks in development: The case study of a wireless project in Nepal. *The Electronic Journal of Information Systems in Developing Countries*, *49*(1), 1–16.

Uden, L. (2007). Activity Theory for designing mobile learning. *International Journal Mobile Learning and Organisation*, *1*(1), 81–102.

Walsham, G. (2006). Doing interpretive research. *European Journal of Information Systems*, *15*(3), 320–330.

Yamgata-Lynch, L.C. (2010). *Activity systems analysis methods: Understanding complex learning environments.* London: Springer Science.

Zins, C. (2007). Conceptual approaches for defining data, information, and knowledge. *Journal of the Association for Information Science and Technology*, *58*(4), 479–493.

5 Actor-network theory

Table of contents

Introduction

This chapter is divided into six main sections in a sequential order. The chapter is introduced in the first section. It begins with a brief description of information systems (IS) and technologies. This is followed by a discussion on ANT from the perspective of IS research. The strengths and limitations of ANT are explicitly discussed, while ANT and its value to IS research are respectively discussed in the sections that follows. Thereafter, is a section that covers ANT and IS research. In the next section, an approach on how to apply ANT in IS research is presented, followed by the conclusion.

Due to the significance of information technology (IT) solutions, many organisations including employees and management wholly rely on them. Organisations make use of IT solutions to support their operations, administration, business processes, and competitive advantage. The significance associated with IT solutions draws keen interest to their deployment (development and implementation), selection, management, or use in organisations. Intensive efforts reveal that the development and implementation of IT solutions are not as easy as individuals and groups are led to believe.

DOI: 10.4324/9781003184119-5

Also, even though it is intended to address challenges, it could by itself be challenging to deploy in various ways.

Technology by itself does not make up an IT solution. There are many factors, including human interaction, that are involved in enabling or transiting technology into solutions. The interplay by humans is critically important in the deployment and governance of IT solutions. The interaction or interplay between humans on the one hand and between humans and technology on the other can be complex and this can increase the already complex nature of IT solutions. The complexity of the interactions comes from the fact that there is no formula or standardised way that underpins them, leading to conventional or unprecedented and unintended circumstances. This has been instrumental to the success and failure of many IT solutions in organisations. Many approaches have been employed by both business and academic domains to assess or evaluate the nature of the interactions, how they are applied, and their consequences; yet the deployment, management, and use of IT solutions continue to be problematic in many organisations.

Owing to the negative effects that the challenges have on goals and objectives, including competitiveness and sustainability, many studies have applied actor-network theory (ANT) as a lens to assess the numerous challenges that are associated with various aspects of IT solutions. The interaction that happens between humans and non-humans (IT solutions and involving processes), which ANT allows us to methodologically examine, has been helpful in IS research. Bowker and Star (1996) suggested that treating classifications and standards with ANT allows political and ethical issues to be addressed, in part, by making infrastructure non-transparent. Although the use of ANT significantly adds value and contributions from theoretical, methodological, and practical perspectives to IS research, there is another momentous challenge, which is the lack of documented way or ways on how to apply the theory.

In addition, the unpredictable nature of business and rapidly changing user requirements make it even more difficult to develop and implement systems in organisations. Other challenges are within the social context, such as politics and culture affiliations. Through the lens of ANT, the understanding of the social context of how IT solutions are developed and implemented is gained. Although ANT has been employed in many studies, it is significant to establish and clarify the factors, from the social perspective, that influence the development, implementation, or management of IT solutions in organisations. ANT aims to understand the processes that lead to the construction and transformation of sociotechnical networks. In the chapter, it is emphasised in the interest of other researchers for the purposes of analysis that no one acts alone as there are important roles played by different actors.

The lack of a documented approach brings about three crucial factors: (1) it slows the popularity of ANT and makes it look far more complicated than what it really is; (2) it is not applied where its contributions have been most valuable, particularly in postgraduates' research; and (3) the balance between human and technology including processes remains far reaching and continue to create detrimental gap in the deployment of IT solutions. In this chapter, I acknowledges the strengths and limitations of ANT but focuses on appreciating and valuing what the theory has added to IS research in solving and finding solutions to the challenges that IT solutions pose to organisations. However, the chapter presents a distinctive argument that the factors that have been identified are not necessarily limitations, but subjective views in applying the theory in IS research. The factors guide and bring fresh perspectives in the course of data analysis, using ANT as a lens. It helps researchers to avoid pitfalls, whether from the criticism or from the strength perspective, in applying ANT to IS research.

Also, ANT provides a platform which allows for the analysis of both human and non-human interaction in a network. Thus, it can critically assess problematic sets of explicit and implicit assumptions made from the human perspective on non-humans, and vice versa. The platform that it provides is a fresh perspective on the importance of relationships between humans and non-humans for both macro and micro structures, and from one progression to the other. It thus focuses attention on the sociotechnical networks that humans create to accomplish individual or group goals and interests through a symmetric value. Another intricate challenge is the use of power, gained from various sources, to influence and manipulate decisions and activities in the deployment and governance of IT solutions in organisations. Within an actor-network, humans or non-humans acquire power to consciously or unconsciously make a difference through the extensiveness and stability of the connections routed around and through them.

Both human and non-human actors work together as a collective to deliver information system as requested by an organisation. Hence, insights into the processes and activities involved in the development and implementation of IT solutions are essentially useful. ANT integrates both human and non-human actors to form or create a network which substantially helps to distil the relationship and interaction that shape and influence IT solutions in an organisation. The main purpose of having a network is to allow actors from various areas to collaborate in order to solve a problem, such as the development and implementation of IT solutions in an organisation. A network is a formation of a group of actors who work together towards a common interest. An employee can participate in many networks and perform different tasks, including making decisions, a scenario defined as heterogeneity in ANT. According to Latour (1996), each element in the network is simply defined by the heterogeneous list of its associates. Also, entities are defined through their existence and their specific associations with each other, and they are not substitutable.

Actor-network theory

ANT is considered a theoretical or a methodological approach, or both, primarily because it has its own epistemological and ontological positions. From an ANT perspective, everything in the social and natural world exists in constantly shifting negotiations in networks of relationships. It thus focuses on the evolution or shifting network-based relationships. For slightly different reason, Walsham (1997) described ANT as both a theory and a methodology primarily because it allows tracing and documentation of (i) network elements; (ii) processes of translation and inscription; (iii) the creation of black boxes or immutable mobiles; and (iv) the degree of stability and irreversibility of networks and their elements. The main tenets of ANT are briefly described in Table 5.1. In addition to the descriptions, each of the tenets is further discussed relative to IS research.

Actor – An actor is any entity that is able to connect texts, humans, money, etc., in order to build more or less effectively a world that is filled with other entities having their own history, identity, and relations (Callon, 1991). In essence, actors are any entity, human and non-human, that have the ability to make a difference, and a network is constituted of actors that are aligned with a common interest. Actors and their interactions together form a network, in any circumstance, at any place, and over a period.

Network – According to Gilding and Tatnall (2005), the formation of a network is not the end of the story, as networks can be unreliable and

Table 5.1 Tenets of an actor network

Tenets	*Description*
Actor (or actant)	Both human beings and non-human actors such as technological artefacts are actors.
Actor network	A group of actors with aligned interests. A network could consist of people, organisations, computers, and standards.
Black box	A state where little or nothing is known about the success of a stable network. A frozen network element, often with properties of irreversibility.
Immutable mobile	Network element with strong properties of irreversibility and effects that transcend time and space.
Inscription	Translation of an interest, which actors represent or speak on behalf of, to express particular viewpoints that have been inscribed in them by the focal actor.
Irreversibility	The degree to which it is subsequently impossible to go back to a point where alternative possibilities exist.
OPP	A narrow path that forces actors to converge on a certain topic, action, or purpose. It is thus an essential element for the formation of a network.
Translation	Creating a body of allies, human and non-human, through a process of translating their interests to be aligned with the actor network.

become unstable. An actor-network that is known and predictable in a certain situation and context is said to assimilate into a black box (Law, 1992). ANT is uniquely concerned with how humans and objects are brought together in stable, heterogeneous networks of aligned interests through processes of translations and negotiations (Callon, 1986; Law, 1991). As pointed out by Iyamu (2013) and Latour (1996), heterogeneous networks overcome issues related to identity and avoid arbitrary dichotomies and structures. Heterogeneity reflects in different organisational principles that are in simultaneous action and interaction.

From the perspective of ANT, a stable actor-network is labelled as a black box so its analysis may focus only on its inputs and outputs (Law, 1992). The stability of a black box is influenced by the materials that are involved in the network. Latour (1996) has since argued that, thus, ANT networks incorporate both human and non-human actors, with the linkages consisting of stabilised translations and interactions between actors.

Inscription – It is the act of proposing a pattern of use as the solution, followed by the attempt to enforce it on others. According to Lero, Lero, and Whelan (2012), inscription translates specific interests within technical objects, such as text, software, and user requirements, which typically have impact on actors' roles. Whether inscription is followed or avoided depends on the irreversibility of the organisational rules and regulations. Inscriptions may lead to irreversibility, which refers to both the degree to which in a certain situation it is impossible to go back to a point where alternative possibilities exist and the extent to which it shapes and determines future translations (Callon, 1991; Hanseth & Monteiro, 1996). A materialised translation, hence mobile, that can be interpreted in essentially the same way in a variety of contexts (i.e., relatively stable in space and time) is referred to as an immutable mobile. Such immutable mobile entities often possess strong properties of irreversibility, e.g., IT architecture standards regarded as non-negotiable rules in some computing environments. Translation is therefore shaped by the enrolment of actors, as translation influences actions of actors in a network.

Irreversibility – A measure of how difficult it is to undo decisions, and the extent to which these determine subsequent actions. According to Iyamu (2011), the degree of irreversibility of a network may be regarded as a process of institutionalisation, which could happen in two different ways: (i) an increased degree of irreversibility is signalled by a firmer institutionalisation, and (ii) the construction of institutions functions as a way to align the network and make it increasingly irreversible. Callon (1991) emphasised that the degree of irreversibility depends on two primary mechanisms: (i) the extent to which it is subsequently impossible to go back to a point where that translation was only one amongst others; and (ii) the extent to which it shapes and determines subsequent translations.

Translation – It implies transformation, which refers to how actors engage with each other to generate ordering effects (Callon, 1986; Law, 1992). Using

ANT as a lens to guide their study, Andrade and Urquhart (2010) argued that the translation process enables the trajectory of events to be revealed. Callon (1991) emphasises that translation goes beyond the traditional definition of action as it deals with mutual definition and inscription. Actors negotiate or manoeuvre others' interests to align with their own purposely to enrol the actors into his or her network. Cordella's (2010) assessment is that the actors that interplay within a network negotiate their forces through a process known as *translation*. When such translations get embodied in a medium or material, they are referred to as inscriptions (Akrich, 1992). Such inscriptions prescribe a programme of action for other actors, although they can vary in strength and flexibility toward achieving a goal (Hanseth & Monteiro, 1996).

The continuous advancement of social activities and technology as a whole forces individuals to adjust or, otherwise, they are left behind in various circumstances. Ironically, applying ANT has proven to be useful in assessing and evaluating the relationship that exists between social and technical factors that are involved in the development, implementation, management, or use of IT solutions. Cho, Mathiassen, and Nilsson (2008) argued that ANT is regarded as an approach for studying how different actors come together and relate within a bigger network.

Moments of translation

Different actors are involved in the development, implementation, management, governance, or use of IT solutions in organisations. The actors' involvement in the activities and processes of IT solutions inevitably brings about negotiation, in order to reach a common goal in the interest of an organisation.

The moments of translation are often applied as a lens in IS research. Substantially, this is primarily because it fortifies the processes of negotiation, which are highly significant in the life of each IT solution in any environment. The moments of translation consist of four stages – problematisation, interessement, enrolment, and mobilisation – and focus on interaction, negotiation, and transformation of events. In Figure 5.1, the moments of translation are illustrated and thereafter a brief discussion is provided.

i Problematisation. This is the first moment where an actor-network can be formed to solve a problem or to take advantage of a new opportunity. At this stage, the focal actor is responsible for identifying problems for relevant actors to solve (Mahring et al., 2004). The focal actor has to be influential enough to convince others on why the issue has to be problematised.

ii Interessement. This is the process whereby negotiations take place between the main or focal actors and other actors within the network

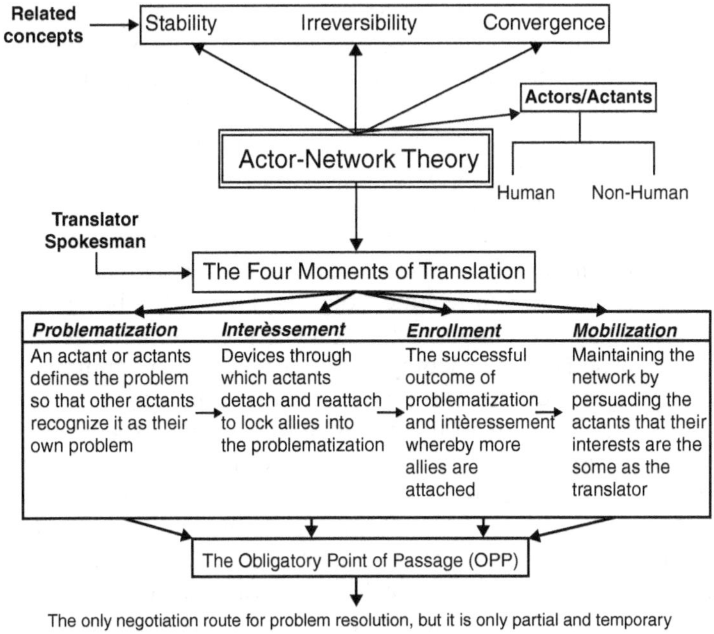

Figure 5.1 Moments of translation (Rhodes, 2009).
Source: Rhodes (2009).

(Iyamu & Roode, 2010). The negotiation determines the interest of the other actors. The roles are defined for the actors who have joined the network. Once the primary actor has negotiated with the other actors, then the next stage is to enrol them to join the network by virtue of executing the tasks that have been assigned to them.

iii Enrolment. Only the actors who accept the roles defined for them in the previous stage participate in the activities of the network (Macome, 2008). Actors are expected to fulfil tasks which are assigned to them, as agreed with the focal actor. According to Luoma-aho and Paloviita (2010), communication plays a vital role in the enrolment stage because it clarifies what is to be expected from the actors and what actions need to be executed.

iv Mobilisation. At this stage, spokespersons are appointed or self-nominated to represent other actors on the network's activities (Callon, 1986). A good representation helps to strengthen the network. It is the responsibility of the spokesperson to engage other actors to perform their tasks and to create the relationship that is desired with other actors in the network. Also, the spokesperson imperatively drives to build trust among actors within the network and other supporting networks including the social system.

In addition to the four moments of translation is the obligatory passage point (OPP), which is an entity that is responsible for representing other actors in a way that suits their significance and actions in the world of translation (Law & Callon, 1997). Actors require a representative who acts on their behalf to address problems and communicate issues. It is the responsibility of the focal actor to define the OPP as it is the channel of communication. According to Callon (1991), OPP is the intermediary between the focal actor and other actors by which the focal actor is indispensable.

In an explanation, Iyamu and Roode (2010) emphasised that ANT does not differentiate human from non-human actors or make a clear distinction between technical and non-technical objects. Thus, ANT brings a different dimension and perspective in viewing events, processes, and activities that are involved, whether in the deployment and management or in the use of IT solutions in an environment. Otherwise, the same challenging results would remain and organisations would continue to be stagnant, and swamped in resolving operational issues, rather than innovation and strategic directions. The development or implementation of an IT solution involves many activities, such as planning, analysis, design, development, implementation, and maintenance. Each of these activities is of primary importance and brings a different type of character. As such, the characters require a microscopic view, such as using ANT as a lens to engage and gain a deeper understanding of how the factors or activities manifest, to produce certain outcome.

The activities which are involved in the development or implementation of IT solutions primarily, are centred on three components: people, technology, and process (Iyamu, 2010). ANT covers the three components comprehensively, from either a human or a non-human actor's perspective. However, in ANT, the components are treated equally, with no bias on the actors' individual contribution within a network. This helps create or establish the balance between humans and non-humans' contribution to the success or failure of an IT solution in an organisation.

One of the aims of applying ANT in many studies is to utilise the moments of translation in order to identify the problems encountered during the development or implementation of IT solutions. Iyamu (2010) argued that due to task dependency and its connections, the implementation becomes diverse in such a way that it requires networks of different actors. For this and other reasons, it is therefore essential to understand the challenges, including the disjoint between the development and implementation phases of IS; how the networks are formed; and how the networks can be effectively utilised to solve challenges, from problematisation to mobilisation stages.

Strengths of ANT

ANT's perspective attempts to explain and interpret social and technological evolution, using neither technical-material nor social reductionism. Rather, it incorporates principles that integrate both humans and non-humans into the same conceptual framework. The treatment of both human and non-human actors by ANT is based upon three principles: agnosticism, generalised symmetry, and free association. The principles are explained as follows:

i Agnosticism – this means analytical impartiality towards all actors, whether human or non-human, within the phenomena under examination or consideration (Callon, 1986).
ii Generalised symmetry – this offers to explain the conflicting viewpoints of different actors in the same terms by using an abstract and neutral vocabulary that works the same way for human and non-human actors. According to Tatnall and Gilding (1999), neither the social nor the technical elements in 'heterogeneous networks' should then be given any special explanatory status.
iii Free association – this requires the elimination and abandonment of all a priori distinctions between the technological or natural, and the social aspect (Callon, 1986).

ANT's strengths are based upon the above principles. The strengths are hereby classified into six categories: iterative process, stability, negotiation scheme, process-oriented, power relationship, and systematism of ANT:

i Iterative process – the process of iterative-ness of actors' interactions and actions over a period and within environment. ANT allows items that have gone the full cycle of re-engineering through the moment of translation to be re-problematised.
ii Stability – ANT attempts to answer how a diverse group of actors reaches agreement at all times; that is, how a social order establishes a certain degree of stability or exhibits structural properties. In ANT, stability is the result of the social process aligning an initially diverse collection of interests into an acceptable stability (Callon, 1991).
iii Negotiation scheme – ANT provides a fresh perspective on the importance of relationships between actors, which is acquired through negotiation over time. This perspective is very important in the development and implementation of technology in the organisation. Both the development and the implementation of technology solution require the process of negotiation, interest, and norm (Akrich, 1992).
iv Process-oriented – the relational and process-oriented sociology of ANT provides a solid theoretical underpinning for unravelling and understanding the impact human and non-human elements have on each other (Tatnall & Gilding, 1999).
v Power relationship – the offering of a precise and non-functionalist account of how actors become established as powerful entities through the stability of the networks that pass through them. The actor (human or non-human) that enacts itself as an OPP in a network has 'power' and greater authority over others. The networks that embed OPP increase the 'power' of an actor. As a result, over time, the ability of an actor to act effectively on a larger scale becomes established (Callon, 1991).

vi. Systematism – ANT systematically avoids methodological dualism. This is the drawing of a priori distinction between technical and non-technical elements (Bloomfield & Vurdubakis, 1997). It simplifies the complexity of assumption and pretence of dealing with two separate, but related, ontological domains. The domains are rather regarded as phases of the same essentiality of action in IS research.

In Walsham's (1997) explanation, some IS researchers either elucidate technology at the expense of social interactions or conversely portray social interaction without giving a detailed description of technological inscriptions. Table 5.2 is intended to provide a guide for addressing this concern, by aligning the interaction and connection between human and non-human actors within context. The six elements of ANT strengths, as discussed above, form the connectivity between human and non-human actors, as shown in Table 5.2. The table is primarily to draw the link between human and non-human actors. The link is intended to guide researchers and avoid disconnects in the use of ANT as a lens for data analysis.

Table 5.2 Strengths of an actor network

Strength	*Human*	*Non-human*
Iterative process	This elicits those who initiate and those involved in the different processes of networks to enact interests, objectives, and goals.	Explanation on how a communication scheme (channel and medium included) is applied to allow for continual interactions and actions.
Stability	This includes those who define, enable, and support the stability of a network over time and space.	The facilities that constitute the formation of stability of a network, within which the actors operate to achieve their goals.
Negotiation scheme	This involves humans who create negotiations and participate in negotiation of events towards specific goals and interests.	The tools that were used to carry out negotiation. This should expose how the negotiation was carried out.
Process-oriented	Humans who create and participate in routine events. Exhumes the factors which entice them to do so at a given time.	The factors that were used to formulate the rules and regulations within the network. Reveals how the rules and regulations were applied.
Power relationship	Those who dominate in the network in the course of an event do so based on the resources within their reach and control.	The factors that were used to dominate and exert control over others in the course of events.
Systematism	Those who design and participate in a systematic approach for a specific goal.	The facilities that were used in the systematic approach for a specific goal.

The strengths of ANT, as presented in Table 5.2, are useful components to guide and validate interaction and connection between human and non-human actors in the analysis of data. In interpretive studies, the analysis is enacted by the dependence, reliance, and interaction between human and non-human actors within a network, over time. Through the interactions and actions of both human and non-human actors, networks are formed and shaped. This amplifies the fact that without the existence of either actor, a network disappears.

However, based on the impressive strengths of the theory, from a sociotechnical perspective, it has been extensively used in IS research. The theory embeds rigour into IS research, particularly when applied as a lens in the analysis of qualitative data. Despite the strengths and the wide use of ANT in IS research, there exist limitations in the theory. Some of the limitations of the theory have been used as criticism by different researchers and authors in IS.

Limitations of ANT

With every advantage comes a disadvantage in the sense of naturalism and realism. So is the case with ANT; it has its strengths as well as limitations which have been uncovered and used as criticism of the theory. In his review, Walsham (1997) recognises four broad strands of criticism as follows: (i) its limited analysis of social structures; (ii) its stance on moral and political issues; (iii) its symmetric treatment of humans and non-humans; and (iv) its tendency to describe, as opposed to explain, together with the problem of managing masses of detail.

I do not pretend to assume that these are the only criticisms of ANT. I, however, prefer to focus on the categories as presented by Walsham (1997). This is mainly because of the popularity that the limitations have gathered over the years – Google Scholar indicates that the paper has been cited over 1,000 times. My assessment of the criticisms and limitations is neither for nor against the theory. It is intended to give a better perspective to have a deeper view of the extent of the limitations in terms of using the theory as a lens for analysis and interpretation of data in IS studies. The limitations, as discussed below, are subjectively abstracted within the characteristics of macro and micro structures. The limitations are hugely based on the interplay and interactions between human and non-human actors.

Macro and micro entities connect through the interaction of human and non-human actors at the point of equilibrium. The interaction of actors in the transformation and evolution of entities at both macro and micro levels, from technologies to ontologies perspectives, is decided by the combined roles of human and non-human actions. As shown in Figure 5.2, this includes how reproduction of evolution continues to occur over time at both macro and micro levels; and how humans and non-humans facilitate the evolution of macro and micro entities over a period and within an environment. If the links are to be overlooked or influenced by the superiority of actors, it would

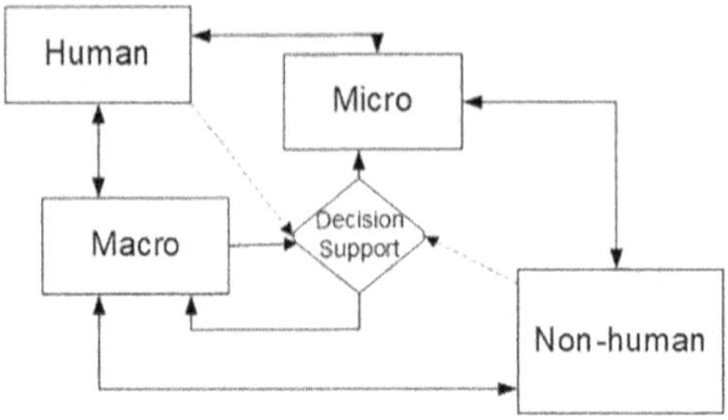

Figure 5.2 Macro and micro connect.
Source: Author.

create other challenges, such as imbalance and disconnect in the evolutionary analysis and interpretation of the characteristics of macro and micro entities within networks.

The macro and micro levels and structures result from both human and non-human actions, through interconnectivity, interaction, or interrelationship. Within a phenomenon, a change at one level affects the other. As a result, the methodology that is used at both macro and micro levels within a network does require a point of equilibrium in its characteristics. Thus, there should not be superiority and imbalance between the actors at either the macro or the micro level. The limitations, as discussed below, should be read in conjunction with Figure 5.2 to gain a better understanding and put things in perspective. The discussion comprises Latour's responses and my summation, together intended to address ANT's limitations as highlighted by Walsham (1997): (i) social structures; (ii) moral and political issues; (iii) symmetric treatment of humans and non-humans; and (iv) the tendency to describe, as opposed to explain.

Social structures

ANT concentrates on how things get done to the detriment of how broader social structures shape socio-material practices. It gives interesting accounts of local contingencies and material arrangements without considering macro-social structures which influence the local ones.

> Actor-network methodology can be used to move between levels of analysis, that the macro-structure is of the same make-up as the micro-structure, and that the macro-structure can be investigated with the same methodological tools as the micro-structure.
>
> (Galliers & Currie, 2011: 307)

As much as this criticism can be considered to be fair, it requires context and perspective. The use of social structure does not necessarily limit or create a vacuum in the analysis and interpretation of data in a sociotechnical context of IS research. It is in prospect, actuates and creates uniformity in the analysis and interpretation of data at both macro and micro levels, in that micro structure and macro structure have relatively the same characteristics, in progression to achieving set goals. Latour (1996) did caution about the differences between an ANT network and technical or social networks. Latour's argument was that, in the concept of ANT, social networks exclude non-human actors in the same way that technical networks exclude the human. Thus, ANT networks incorporate both, with the linkages consisting of stabilised translations and interactions between actors.

Moral and political issues

Another criticism is that ANT destroys the credibility of science by leaving no place for the objective truth that science (allegedly) uncovers. Of course, sociologists deliberately avoid making commitments of this kind (Lee & Brown, 1994).

ANT is said to have disregard for moral and political issues. This disregard affects the macro structure and has led to criticisms of ANT, and of strong social constructivism and relativism in general, for being amoral and apolitical in that it leads to ignoring the political biases that can underlie the spectrum of choice for relevant actors (Winner, 1993: 370). ANT does not provide specific guidelines for examining the ethical and moral implications of IT (Bijker, 1995; Walsham, 1997). Star (1991) refers to the 'networks of the powerful' and talks about how irreversible networks are only stable for some and discriminate against those who do not belong to the community of practice, those who use and maintain the network. Partial response to the criticisms of amoral and apolitical relativism is as follows:

> *Refusing to explain the closure of a controversy by its consequences does not mean that we are indifferent to the possibility of judgments that transcend the situation. For network analysis does not prevent judgment any more than it prevents differentiation. Efficiency, truth, profitability, and interest are simply properties of networks, not of statements. Domination is an effect not a cause. In order to make a diagnosis or a decision about the absurdity, the danger, the amorality, or the unrealism of an innovation, one must first describe the network'.*
>
> (Latour, 1991: 130)

Morality and politicism are entities of subjectivism, which is synonymous individual thought and experience. Subjectivism is a thought objective and conscious state of an individual experience and feeling. As such, it is difficult to provide a formula or guide without bias. Also, it is difficult or impossible to balance and maintain the uniformity of bias within a context and relevance. Otherwise, this creates obscurity in the analysis or interpretation of data relating to macro structure or micro structure.

Symmetric treatment of humans and non-humans

Symmetric treatment is a criticism of ANT because it uses the same language to describe both human and non-human actors. The use of the same language occurs because ANT introduces symmetry between humans and non-humans by anthropomorphising things. This criticism claims dehumanisation of humans due to equal treatment with non-humans.

The symmetry between human and non-human actors, which is related to the symmetry between social and technical, society and nature, politics and science, and values and facts, has been criticised for having gone too far in eliminating all distinctions and reducing humans to the status of things. But how and under what circumstances actants such as computers assume characteristics different from humans is part of the problem to be addressed (Woolgar & Grint, 1991).

> It was emphatically '... to avoid the twin pitfalls of sociologism and technologism. We are never faced with objects or social relations; we are faced with chains which are associations of humans ... and non-humans ... No one has ever seen a social relation by itself ... nor a technical relation'.
>
> (Latour, 1991: 110)

The criticism could be based on the fact that humans are made to compete and equalled with non-humans. The fact is non-human actors such as computers (or IT solution) are created by humans; in essence, humans are competing with themselves for power through their own innovations. Also, lack of equality of both humans and non-humans in the context of data analysis instils superiority of human over non-human actors. The question is, therefore, how does one assess human superiority vis-à-vis non-human superiority in the context of an actor network? The human creates the non-human but cannot live or act without the entities that he or she has created. Without equality in the context of a phenomenon, it becomes difficult or impossible to avoid bias in the data analysis of the interaction that occurs between human and non-human actors.

Its tendency to describe as opposed to explain

Another criticism of ANT is that it fails to provide explanations for the dynamic restructuring of networks. It is in addition argued that ANT fails to take account of the effects that technology can have on those who are not part of the network and that it also fails to support value judgements on the desirability or undesirability of such effects. ANT tends to recognise only the actors that have a role in a network, and not those that are impacted by the networks (Winner, 1993).

> In Latour's (1991: 129) argument, describing the network is not only a prerequisite, but also the only way to get explanations: '*The explanation emerges once the description is saturated; if we display socio-network-defining trajectories by actants' association and substitution defining actants by all the trajectories in which they enter, by following translations and finally, by varying the observer's point of view – we have no need to look for additional causes. Explanation is the stabilization of a network. If one is capable of explaining effects of causes, it is because a stabilized network is already in place*'.

Explanation can be definitive, which makes it difficult or inflexible to accommodate other entities, which might be relevant. Within a network, an actor can be anything as long as it makes a difference. Within that context, it is difficult to be definite about explanation. Also, an explanation is subjective and applies to context and relevance in the course of natural settings, as in the case of macro structure and micro structure. Thus, an explanation can only be applied as the network is followed in the examination and understanding of the interactions amongst the actors through a narrative process. Otherwise, the explanation would be based on speculation and short of substantive e. According to Callon (1991), explanations are offered only by networks which increase their convergence and irreversibility, and this is when the description delivered by intermediaries turns into explanations (and even predictions).

Applying ANT in IS research

To apply ANT in IS research, it is essential to understand how. First, understand the strengths and limitations of ANT as covered above, which are intended to create awareness and provide a guide for researchers in applying the theory in their IS research. This section continues along the same line, but focuses on a symmetric perspective in the use of the theory in IS studies. Second, understand that ANT can be applied at both macro and micro levels in analysis. This is primarily because ANT does not only focus on the creation of networks, but it also helps to identify the roles, technologies, interactions, and interrelations that exist and influence action and activities in the deployment, management, and use of IT solutions in an organisation.

Despite the criticisms, ANT is useful conceptually, and as a lens for assessing and evaluating the complex reality that regularly confront IT solutions from both human and non-human perspectives. As shown in Figure 5.3, the non-human artefacts such as technology are as a result of human actions through the use of facilities in heterogeneous networks. The

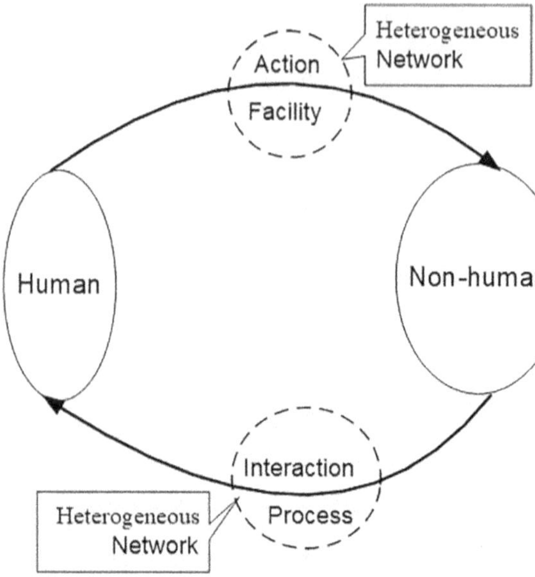

Figure 5.3 Human and non-human symmetric interplay.
Source: Author.

same non-human artefacts that are created by human actions are used to enable and constrain human activities through interactions and processes, at both macro and micro levels, over a period and within an environment. Thus, it is difficult or impossible to dissociate humans from non-humans within a phenomenon, or to begin to evoke superiority over what could be a determining factor to the existence of an action. Figure 5.3 depicts facility enablement of actions and, in response, processes that facilitate interactions between humans and non-humans in a heterogeneous network.

Symmetrically, both the strengths and limitations that have been identified in ANT are of significant value in the course of data analysis and interpretation, through the balance that they create between human and non-human actors. The limitations, as discussed in the preceding section, do not necessarily create a gap in applying ANT as a lens either in analysis or in interpretation of data in IS research. In illustrating the non-real effect, I employ an analogy to an argumentative stance of the symmetrical value of the equality of all actors within a network: *all fingers are claimed to be equal by the social justice of the land, hence the formulation and promulgation of one common Law or Constitution for all that live in it. Even though we undeniably know that some fingers are longer than the others, and that they can never be equal, all the fingers are governed by but a common law.* If humans and non-humans are associated with unequal symmetric values, it becomes skewed to measure. As a result, their combined contributions make a difference within a network, thereby affecting the analysis of interactions between the actors.

The coexistence and inseparability of human and non-human actors to produce and reproduce activities make it difficult to symmetrically treat them unequally. A discriminative approach would make it impossible for actors to be able to translate the capacity and capability to make a difference in the transformative process of structures in a social system. Also, apoliticism from an ANT perspective helps to engage and balance the roles of both human and non-human actors within a network of social activities. The equal explanation for both humans and non-humans provides a prescriptive guidance for the identification of different interests and values within a network, through which a black box can be created for entities, such as macro and micro structures.

ANT as a lens

As has been established in this chapter, the events and activities concerning IT solutions are of both technical and non-technical nature. The roles of each actor are considered important, primarily because individual action brings a difference to the success or failure of an initiative. The deployment or governance of IT solutions is more challenging than is often portrayed. This could be attributed to the fact that a number of unique skilled personnel, processes, activities, and other factors involved in IT solutions' attainment are not deeply known, to the extent that they require specific attention. Every individual within a team is assigned tasks scheduled to be completed within a timeframe.

Another goal of ANT is to create a diverse network which consists of humans and objects with related interests through the moments of translation of the events. The moments of translation involves four stages, namely, problematisation, interessement, enrolment, and mobilisation. It is through these stages that a diverse network is formed, and information is passed to those who require it. In the moments of translation, the roles and responsibilities of actors that are involved in the network are explained. Through the different moments, information is translated in achieving the common interest of the network.

The developments or implementations of IT solutions are joint efforts, as they involve many participants in their social and technical contexts, as depicted in Figure 5.4. Participants are also known as actors; they can be either people or technologies. What remains unclear and is hard to diagnose ordinarily is the actual capability and predictive actions of the actors, which occur at any given time. This is the strength of ANT as a lens as it can be used to look deeper than at mere speculations by employing Figure 5.4. The constant shifting of negotiation, as depicted in Figure 5.4, is presented as a preparatory foundation in applying moments of translation (Figure 5.5) for the analysis or interpretation of data in IS research.

Humans' primary involvement in the deployment, management, or governance of IT solutions is to render services, whilst technologies are utilised

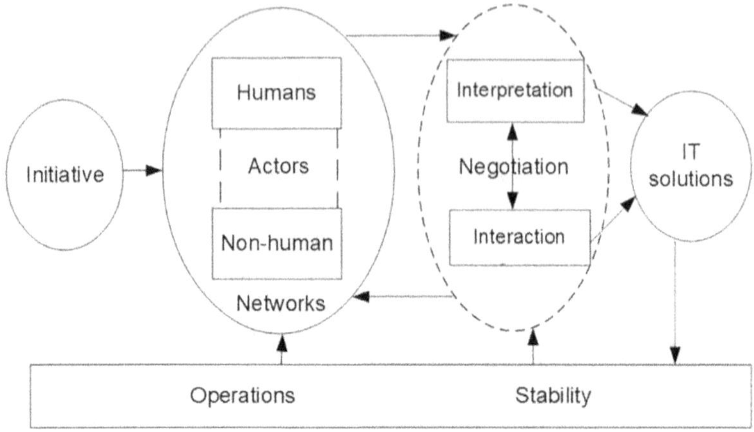

Figure 5.4 Shifting negotiation of actor networks.
Source: Author.

to activate, enable, and support processes and activities in an organisation. In the course of doing so, both human and non-human actors provide equal support or service to the network. The identification of actors and their roles or responsibilities prior to carrying out the activities concerning IT solutions helps the actors to connect, interact, and depend on each other. Hence, who, what, how, and why questions are essential in every action or activity, helping to critically facilitate analysis or interpretation of data, as in Figure 5.5.

The interdependency between actors (human and non-human) for a common interest and goal within a network facilitates activities to provide a desired solution to a problematised issue. It is thus essential to characterise actors by their actions rather than the positions that they occupy within a structure. Simply identifying both human and non-human elements as actors removes individualisation and fosters cooperation and dependence amongst actors within a network. However, it is a well-known and documented fact that actors either compete or are used to compete against each other. Nonetheless, the intention is usually for them to coexist, collaborate, or complement each other with either the skills that they possess or the weaponry of their existence.

Figure 5.5 helps to affirm ANT's usefulness in the field of IS to address as well foster the interplay of both social and technical elements of an initiative within a network over a period. The nature of the moments of translation ensures an iterative process, which examines the production and reproduction of actions within a system's network. For example, an implemented IT solution can be problematised in full (or in part), and thereby the moments of the translation process re-enacted.

The actors within a network play the roles which support both social and technical factors in deploying, enabling, maintaining, or supporting

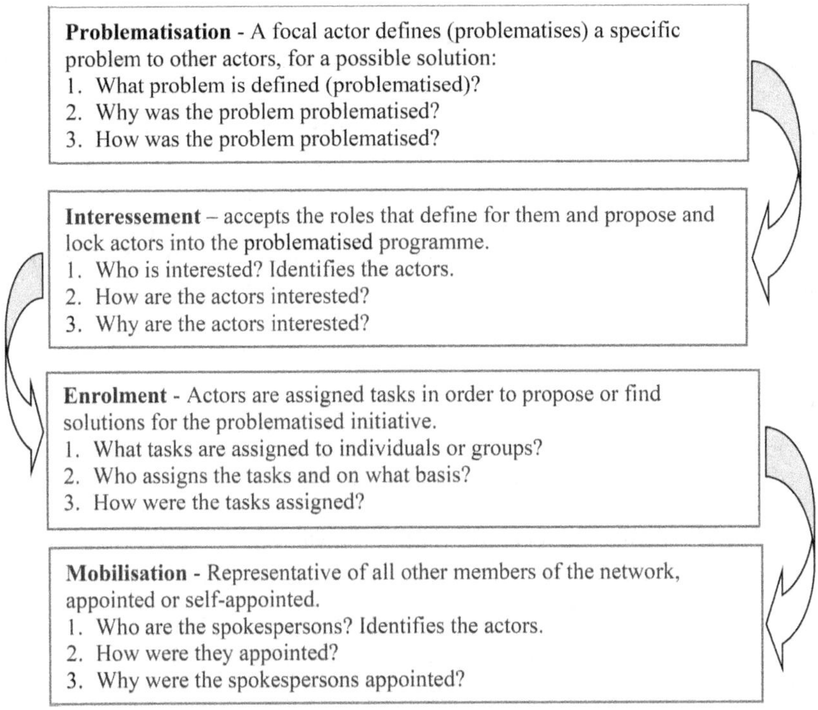

Figure 5.5 Moments of translation as a lens.

IT solutions. Thus, ANT is useful in assessing and evaluating sociotechnical factors influencing IT solutions in organisations. From development to implementation, including use, activities of IT solutions are complex tasks to accomplish. Consequently, the activities require different skill sets, processes, procedures, and enabling plans.

Thus, IT solutions should be engaged as a portion of a dynamic social world rather than simple tools designed to solve social conflicts. This is primarily because technologies are constantly changing for the purpose of making life, processes, and activities easier, better, and more flexible. Analysis and interpretation of data from an IS research perspective are instrumental to ease off the activities. Fundamentally, it requires a vibrant network consisting of human and non-human actors with aligned interests, as opposed to against each other, to comprehend the utilisation of available technologies. Profoundly, the use of IT solutions plays an essential role in solving problems and supporting decision-making in various aspects of organisations. This indicates that organisations and individuals are partly, sometimes wholly, dependent on technology in order to function.

Conclusion

ANT is unbiased and equally supports the roles and involvement of both human and non-human actors. Its use to evaluate activities is essentially important. Applying ANT in IS research does not only focus on the creation of networks, but it also helps to identify the roles, technologies, and the connection between the two – human and non-human actors – within networks. Due to much interdependency between the actors, and the influential nature of some actors, it is vital to consider employing OPP at all times. Also, it is essential for the focal actor to have negotiation skills. In the activities concerning IT solutions, moments of translation vitally facilitate negotiations.

References

Akrich, M. (1992). The description of technical objects, In: W. E. Bijker & J. Law (eds.), *Shaping technology, building society*, pp. 205–224. Cambridge: MIT Press.

Andrade, A., & Urquhart, C. (2010). The affordances of actor network theory in ICT for development research, *Information Technology & People*, *23*(4), 352–374.

Bijker, W. E. (1995). *Of bicycles, Bakelites, and Bulbs: Towards a theory of sociotechnical change*. Cambridge: MIT Press.

Bloomfield, B., & Vurdubakis, T. (1997). Visions of organization and organizations of vision. *Accounting, Organizations and Society*, *22*(7), 639–668.

Bowker, G., & Star, S. (1996). How things (actor-net) work: Classification, magic and the ubiquity of standards. *Philosophia*, *25*(4), 195–220.

Callon, M. (1986). Some elements of the sociology of translation: Domestication of the scallops and the fisherman of St Brieuc Bay. In: J. Law (ed.), *A new sociology of knowledge, power, action and belief*, pp. 196–233. London: Routledge.

Callon, M. (1991). Techno-economic networks and irreversibility, In: J. Law (ed.), *A sociology of monsters. Essays on power, technology and domination*, pp. 132–164. London: Routledge.

Cho, S., Mathiassen, L., & Nilsson, A. (2008). Contextual dynamics during health information systems implementation: An event-based actor-network approach. *European Journal of Information Systems*, *17*(1), 614–630.

Cordella, A. (2010). Information infrastructure an actor-network perspective. *International Journal of Actor-Network Theory and Technological Innovation*, *2*(1), 72–53.

Galliers, R. D., & Currie, W. (Eds.). (2011). *The Oxford handbook of management information systems: Critical perspectives and new directions*. New York: Oxford University Press.

Gilding, A., & Tatnall, A. (2005). Actor-network theory in information systems research. *Encyclopaedia of Information Science and Technology, 1*, 42–46.

Hanseth, O., & Monteiro, E. (1996). Developing Information Infrastructure: The tension between standardization and flexibility. *Science, Technology & Human Values*, *21*(4), 407–426.

Iyamu, T. (2010). Theoretical analysis of strategic implementation of enterprise architecture. *International Journal of Actor-Network Theory and Technological Innovation*, *2*(3), 17–32.

Iyamu, T. (2011). Institutionalisation of the enterprise architecture: The actor-network perspective. *International Journal of Actor-Network Theory and Technological Innovation*, *3*(1), 27–38.

Iyamu, T. (2013) Underpinning Theories: Order-of-use in information systems research. *Journal of Systems and Information Technology, 15*(3), 1–13.

Iyamu, T., & Roode, D. 2010. The use of structuration theory and actor network theory for analysis: Case study of a financial institution in South Africa. *International Journal of Actor Network Theory and Technological Innovation*, *2*(1), 1–26.

Latour, B. (1991). Technology is society made durable, In: J. Law (eds.), *A sociology of monsters: Essays on power, technology and domination*, pp. 103–131. London: Routledge.

Latour, B. (1996). Social theory and the study of computerised work sites. In: W. J. Orlikowski, G. Walsham, M. R. Jones, & J. I. DeGross (eds.), *Information technology and changes in organizational work*, pp. 295–307. London: Chapman & Hall.

Law, J. (1991). Monsters, machines and sociotechnical relations, In: J. Law (ed.), *A sociology of monsters: Essays on power, technology and domination*, pp. 1–23, London: Routledge.

Law, J. (1992). Notes on the theory of the actor-network: Ordering, strategy, and heterogeneity. *Systems Practise, 5*(4), 379–393.

Law, J., & Callon, M. (1997). The life and death of an aircraft: a network analysis of technical change. In: W. E. Bijker & J. Law (eds.), *Shaping technology/building society: Studies in sociotechnical change*, pp. 21–52. Cambridge: MIT Press.

Lee, N., & Brown, S. D. (1994). Otherness and the actor network. *American Behavioural Scientist*, *37*(6), 779–791.

Lero, N., Lero, T., & Whelan, E. (2012). Service science: An actor-network theory approach. *International Journal of Actor-Network Theory and Technological Innovation*, *4*(3), 51–69.

Luoma-Aho, V., & Paloviita, A. 2010. Actor-networking stakeholder theory for today's corporate communications. *Corporate Communications: An International Journal*, *15*(1), 49–67.

Macome, E. (2008). On Implementation of an information system in the Mozambican context: The EDM case viewed through ANT lenses. *Information Technology for Development*, *14*(2), 154–170.

Mahring, M., Holmstrom, J., Keil, M., & Montealegre, R. (2004). Trojan actor-networks and swift translation. *Information Technology & People*, *17*(2), 210–238.

Rhodes, J. (2009). Using actor-network theory to trace an ICT (telecenter) implementation trajectory in an African women's micro-enterprise development organization. *Information Technologies & International Development*, *5*(3), 1–20.

Star, S. L. (1991). Power, technologies and the phenomenology of standards. On being allergic to onions. In: J. Law, (ed.), *A sociology of monsters: Essays on power, technology and domination,* pp. 27–57. London: Routledge.

Tatnall, A., & Gilding, A. (1999). Actor-network theory and information systems research, In: *Proceeding of the 10th Australian Conference on Information Systems*, (ACIS), Wellington, Victoria University of Wellington.

Walsham, G. (1997). Actor network theory and IS research: Current status and future prospects. In: A. S. Lee, J. Liebenau, & J. I. DeGross, (eds.), *Information systems and qualitative research,* pp. 466–480. London: Chapman & Hall.

Winner, L. (1993). Upon opening the black box and finding it empty: Social constructivism and the philosophy of technology. *Science, Technology & Human Values*, *18*(3), 362–378.

Woolgar, S., & Grint, K. (1991). Computers and the transformation of social analysis. *Science, Technology, & Human Values*, *16*(3), 368–378.

6 Contingency theory

Table of contents

Introduction

In recent years, many sociotechnical theories have been introduced and applied to underpin information systems (IS) studies, particularly in the qualitative paradigm. The increasing interest and attention is mainly because of the shift from simply applying the interpretivist approach in the analysis of qualitative data to a more rigorous approach of applying theories. There are two main research methods, qualitative and quantitative. However, both methods can be complementarily applied in the same study, which is referred to as mixed methods. In this chapter, I explicitly focus on qualitative research methods. Qualitative research explores attitudes, behaviours, and experiences through such methods as interviews, observation, or focus groups (Sekaran, 2003). Its purpose is to attempt to get an in-depth opinion from participants. The qualitative method allows the researcher to ask respondents complex questions, as well as instantly probe responses (Holliday, 2007). This approach has the advantage of allowing for more diversity in responses as well as the capacity to adapt to new developments or issues during the research process itself (Denzin & Lincoln, 2000). Qualitative data is interpretively analysed, primarily because of its 'subjectivist' nature.

Interpretivism focuses on gaining an understanding of the human world, based on real-life experiences. It has been emphasised that 'reality is socially

DOI: 10.4324/9781003184119-6

constructed' (Mertens, 2005). Thus, interpretivist researchers tend to rely upon views and opinions of participants about the situation they experience and around them relative to the phenomenon being studied (Creswell, 2003). As the research evolves, it is somehow no longer so much of benefit and significance to rely wholly on an interpretivist approach as a single entity for data analysis. Otherwise, many analyses will continue to be horizontal and overzealous, as opposed to succinct and vertical, and deeply rooted in sense making and outcome. In recent years, IS studies have been increasingly underpinned by sociotechnical theories, such as the contingency theory (CT). The theory is most useful, particularly in areas such as contingencies and organisational performance enhancement, which in other theories play little or no part in coverage.

CT attempts to address the interrelationships within organisational systems or subsystems as well as an environment. Donaldson (2001) defines CT as a 'variable that moderates the effect of organisational characteristics on organisational performance'. Betts (2003) argues that the CT is optimal, depending on the nature of the environment associated with an organisation. Primarily, CT is concerned with the role of structure in organisational performance. From the perspective of CT, Blanton, Watson, and Moody (1992) developed the contingency model, which consists of environment, structure, performance, and strategy, as shown in Figure 6.1. The social environment encourages employees to associate value with their jobs and perform in such a way that the organisation is able to operate strategically, tactically, and spontaneously. According to Husted (2000), the contingency model permits strategies and structures to be aligned with the organisation's objectives to achieve enhanced performance.

Through this understanding, the theory attempts to address the overall success or failure of an organisation in relation to performance. The efficacy of an organisation results from the goals that are set by managers. In CT, a relationship exists between structure and strategy as emanating from

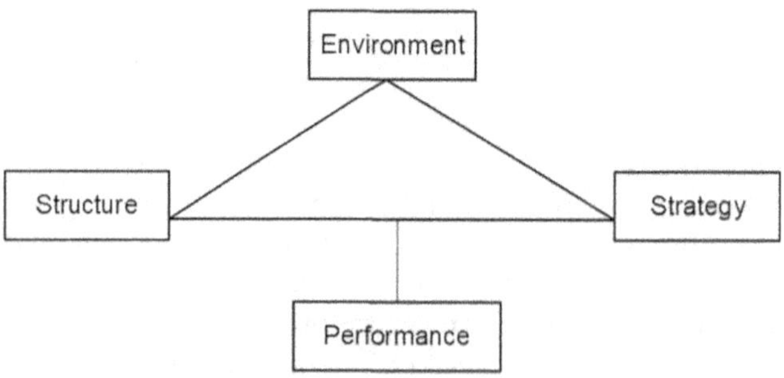

Figure 6.1 Contingency model (Blanton et al., 1992).
Source: Blanton et al. (1992).

the environment in which it was intended to improve on performance. This association should increase the value and provide commonality. By its nature, CT avoids static rules. Thus, the theory holds that a better fit between the organisation and structure has a positive effect on performance that, in turn, predicts favourable outcomes. Contingency dictates that organisational structures change over time and space.

This chapter presents a guide on how CT can be applied to examine, investigate, and understand qualitative data in IS research. The remainder of this chapter is divided into five main sections. The first section presents a review of literature on CT. The second section covers how CT relates to IS studies. The third section presents the underlining guideline, a preparatory format in the application of CT. The fourth section presents and discusses a guide, which can be followed in data analysis, when CT is employed to underpin IS studies. Finally, a conclusion is drawn.

CT overview

CT is based on four main ideas, as proposed by Fred Fiedler in the 1960s (Fiedler, 1964). It is articulated into four categories as follows:

> (i) There is no universal or one best way to manage; (ii) The design of an organisation and its subsystems must 'fit' with the environment; (iii) Effective organisations not only have a proper 'fit' with the environment, but also between the subsystems; and (iv) The needs of an organisation are better satisfied when it is properly designed and management style is appropriate both to the tasks undertaken and the nature of the workgroup.

However, there have been some questions about the word 'fit', understandably so because of the subjectivism that is easily associated with it. The main questions were in the areas of structures, interactions, and functionality. See Drazin and Van de Ven (1985), where clarity has been provided for the questions and the use of 'fit', which seems to have settled the criticisms.

The theory is based on the notion that a fit between certain components of contingencies helps to improve an organisation's performance. As shown in Figure 6.1, contingencies include structure, environment, strategy, and performance. CT assumes that each of these contingencies necessitates the existence of certain characteristics in an organisational structure. Thus, when the structure of an organisation bears those characteristics for the contingency in question, this is an indication that there is a fit, which then promotes and increases the performance of the organisation. The components of the contingency imply a proper fit between organisational context and structure in order for the organisation to succeed or fail in its performances. Many organisations do not only have a proper fit with their environment, but they also have a proper fit among its subsystems. The fit within

CT is critical, as it determines organisational performance, which is vital for growth, sustainability, and competitiveness.

There are internal and external demands that must fit the environment for better performance. Performance is a process which interacts with the relationship that exists between environment and strategy and this, in turn, encourages organisational structure (Blanton et al., 1992). The best organisational structure is the one that fits the organisational information requirements and yields growth. CT helps to assess or evaluate factors that fit positively with performance. The components of contingency are described in the section that follows.

Key elements of CT

CT is concerned with the overall success or failure of an organisation relative to performance. Structural CT is effective in an organisation if it fits its characteristics and the environment, such as its size or strategy. The efficacy of organisational results emanates from the goals and objectives that are set by managers. Structural contingency consists of three main components (Donaldson, 2006), discussed as follows:

i The need for a positive relationship between contingency and organisational structure

The relationship between contingency and structure forms an association that increases value and provides commonality across the structural CT. This consequentially leads to a positive association between contingency and structure, which often creates a conducive environment for competitiveness and opportunities to develop. Such suitable climate is enabled and supported by IT solutions towards greater economic performance.

An organisational structure that supports the current strategy achieves organisational goals and objectives, through IT solutions' enablement. Also, a planned structure within organisations increases the chances of sustainable achievement which, in turn, leads to high performance, thereby satisfying its clients, including the fulfilment of both internal and external factors. An imbalance in structure renders some organisations incompetent, and performance rapidly declines.

ii Contingency change causes organisational structural change

Contingency directs changes in organisational structures from time to time. By implication, structural contingency focuses on change from current to desired states. The change in contingency encourages the organisation to adopt a better fit that leads to improved and beneficial performance. Husted (2000) suggests that the social environment plays a critical role in ensuring the survival of an organisation due to constantly changing expectations of stakeholders.

In addition, contingency causes structures to change in an environment. As a result, it influences the current organisational situation, which provokes drastic change and is intended to improve performance. Organisations strive to achieve high performance as stipulated by top management, which are sometimes influenced by social structure. Consequently, they are forced to change their current operations in an attempt to achieve their goals and objectives. Complex environments within organisations change from time to time in order to address different social issues, including factors of influence, from both negative and positive perspectives (Van Looy & Van den Bergh, 2018). This ultimately leads to enhancement of corporate social performance. Change in contingency means organisational processes can restore efficacy (Donaldson, 2001). Changing structures in some organisations is intended to improve effective service delivery to clients and partners.

iii Fit affects performance

CT holds that an improved relationship between organisation and structure has a positive effect on performance that predicts favourable outcomes. Where the structure does not fit well with the organisation, performance results are low, and competitiveness is derailed. Contingency relates to the relationship between fit and performance (Joslin & Müller, 2016). Fit performance is the most common element of CT. The main objective of fit is to enhance the organisation's social performance (Maruping et al., 2019). Fit enhances performance in organisations to deliver excellent services to both internal and external clients. Performance increases and the relationships between contingency, structure, and environment are stabilised. Organisations need to focus more on the goal to be fit to ensure success.

CT and IS

An organisation requires a reliable IS to provide better service to its customers. IS are the means by which people and organisations, utilising technologies, gather, process, store, use, and disseminate information. The logical extension of the contingency approach is that each case is unique. This is similar to organisations' business goals and requirements. From a CT perspective, success or failure of an IS within an organisation is the fit which results from the system design, technology deployment, and the organisation's performance. If this is true, then management can be practised only by intuition and judgement, thereby negating the value of prior knowledge and wisdom. Every case is considered unique until generalisation is employed. As such, each and every case has to be analysed individually, within context. It is therefore critical that each organisation understands the relationship between fit and misfit before adopting an IS (Van de Ven, Ganco & Hinings, 2013).

Conceptually, applying CT to IS research is primarily aimed at the effect of interaction, explaining what is observed about the phenomenon being studied. It is in this context that the theory helps to assess the influence of environmental factors, such as governance, and the relationship between processes and technology solutions. Consequently, it evaluates the link between IT capabilities and performance of an organisation (Stoel & Muhanna, 2009). Many IS studies employ the model from the perspective of CT by Blanton et al. (1992). CT is a behavioural theory that claims that there is no best way to organise or manage organisations and their structures (Van Looy & Van den Bergh, 2018). It is, indeed, to design an organisational structure that can effectively and efficiently handle certainty or uncertainties in an environment.

The theory holds that the most effective organisational structural design is where the structure fits contingencies. However, there are challenges to these beliefs, some of which are theoretical, while some are empirical. The challenges influence and impact development, implementation, and governance of IS and technologies. These challenges can manifest to be enablers or constraints through theoretical and methodological innovations. Both theoretical and methodological innovations constitute opportunities in examining through the lens of the CT its focus on leadership. The optimal organisation or leadership style is contingent upon various internal and external enablers and constraints. According to Fiedler (1964), there is no one best way of organising or leading; an organisational or leadership style that is effective in one situation may not be successful in another. Even though this argument was established very long ago, things remain the same.

CT can change how IS and technologies are managed during development, implementation, or governance because it introduces the concept that no one management approach will work for every scenario or project (Howell, Seidel & Windahl, 2010). Each project that an organisation initiates may benefit from the use of a different management and technical approach or a combination of approaches. This is attributed to the fact that core components, people, process, and technology are different from one project (or stages of the project) to another. Some organisations prefer to adopt a single standard for all or subsequent projects. CT allows organisations to work together by de-emphasising a 'one size fits all' mentality for managing projects (Howell et al., 2010).

The proper fit to an organisation enables the operation of IS to satisfy its client's requirements. According to information technology, solutions provide different benefits to organisations, such as improving productivity, innovation, and cost savings (Maruping et al., 2019). Thus, IT solutions must be aligned with the organisational structure to accomplish the intended benefits. The fit between an organisational structure and IT solutions, including strategy, enhances the performance of any organisation (Blanton et al., 1992). CT is in the category of behavioural theories which maintain that what is best for one organisation might not work for the others. Hence, the approach of the theory focuses on an organisational structure that suits an organisation within a particular environment. CT also explains organisational performance at different levels to enhance or reduce performance.

Applying CT as a lens

The use of CT requires regular communication, vital in the development, implementation, governance, and management of IS and technologies. This is mainly to allow for team members to seek clarifications and share experiences in addressing organisational challenges, goals, and objectives towards improved performance. The use of CT results in variances from strategic or tactical to operational solutions. Also, the concept of fit, from a CT perspective, enables and addresses inconsistency of structures and disconnect in interactions and relationships between actors and components. Joslin and Müller (2016) suggest that CT conceptually stresses and unpacks the interaction effect of structure on performance.

The contingency model from the perspective of CT was adopted from Blanton et al. (1992) as illustrated in Figure 6.2. It supports the assumption of rational behaviour. If a better IS is designed, employees are willing to utilise the system because it is intended to improve their contribution to organisational performance. As shown in Figure 6.2, applying CT for analysis or interpretation of qualitative data in IS research entails connectivity of three elements – CT Tenets, Category (Incidental Artefacts), and Environmental Components – as shown in Figure 6.2. The tenets are structure, environment, strategy, and performance. The environmental components consist of technology, process, and people. The incidental artefacts are what, who, how, why, when, and where, as discussed:

a CT Tenet

The tenets of CT do not operate in a vacuum but interact with other artefacts towards the fulfilment of goals and objectives, as shown in Figure 6.3. The aims and objectives of the tenets are achieved through the enablement of technology applied by people through processes. Also, the interaction between the tenets and other artefacts (components and category) focuses on uniformity, consistency, and empowerment, within context. In organisations, IT solutions are employed to enable employees' effectiveness and efficiency. Also, the solutions empower users as they centralise and support existing services. The components of the contingency have a positive effect on performance; hence it requires IS support.

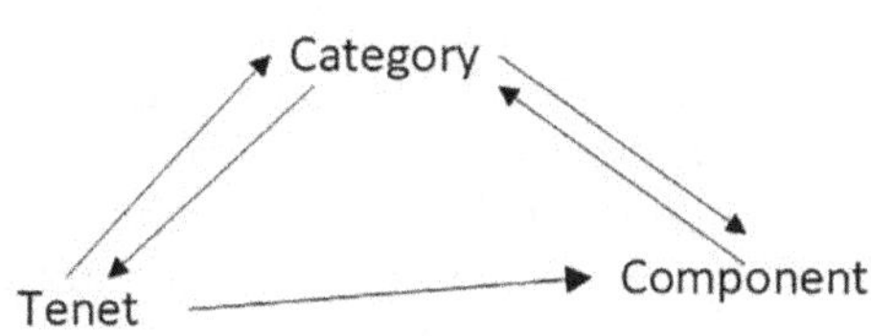

Figure 6.2 Elements in applying CT.
Source: Author.

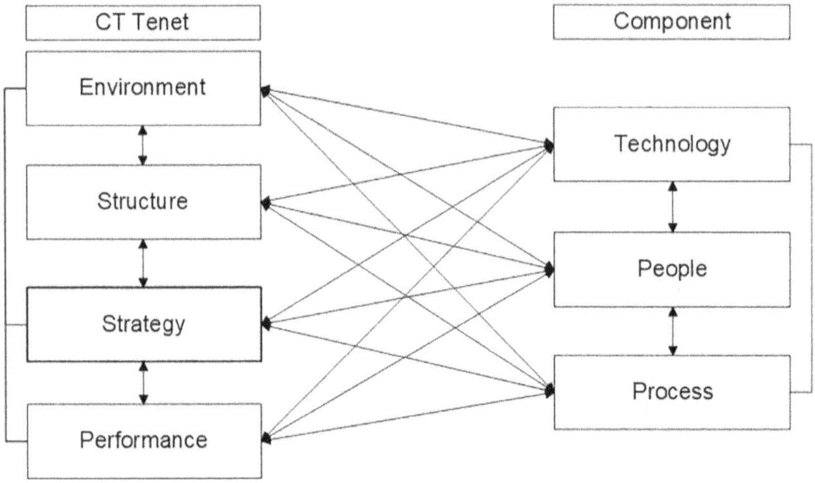

Figure 6.3 Alignment between CT tenets and strategic components.
Source: Author.

The theory focuses primarily on the role of structure in organisational performance. In addition, CT attempts to understand the interrelationships within the organisational structure and environment that cause a harmonious fit to exist (Maruping et al., 2019). The contingency model by Blanton et al. (1992) supports the assumption of rationality in human behaviour. According to Husted (2000), the contingency model of social performance emphasises that strategies and structures have to be aligned with the social issues of the organisation to lead to greater social performance.

b Category

As shown in Figure 6.4, the three categories – technology, process, and people – are in environment, strategy, structure, and performance, in the context of IS and technologies. These components are the key instruments and influential factors in the deployment, management, and support of IT solutions in organisations.

In an assessment by Mooney, Gurbaxani, and Kraemer (1996), many studies that have been carried out that concern process recognise the role of IT in process improvement, but none explicitly considers the business value of IT in process. Hackney, Burn, and Dhillon (2000) argue that a strong process perspective should be adopted before any IT implementation is undertaken. Technology determines the technological solutions needed to enable and support an organisation's objectives, through its IS and technology strategy.

Simply involving people is not enough. The people should be empowered to assume the responsibilities that are required of them to enable and

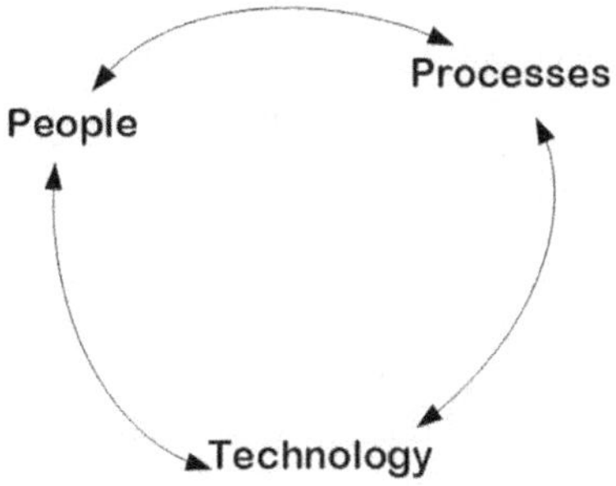

Figure 6.4 Components of an IT strategy (Iyamu & Adelakun, 2008).
Source: Author.

support systems and structures. Using the structurational model of technology, Orlikowski (1992) illustrates that IT is a product of human action within specific contexts. This may require specific training and use of technology through legitimate processes in order for them to apply expertise and contribute through their know-how. People are very influential in determining the success or failure of IT solutions in organisations.

Process is a component that shapes the routes and boundaries within which IS and technologies are developed, implemented, selected, and managed. It engineers stability and normality within an environment. Thus, Giddens (1984) describes it as normative rules which social agents translate as specific rights and obligations. In Davenport and Short (2003), it is argued that process improvement is based on the capability that IT can provide.

c Environmental Component

In order to gain deep insights on why and how things happen in an environment, strategy, structure, and performance, the interaction between individuals and groups with technology and processes are interrogated. This is done through interrogatives such as what, who, why, how, when, and where, to deterministically examine organisations' environment to understand their strategy and structure as they strive to continually improve performance. Zachman (1987) made use of the six interrogatives who?, what?, when?, how?, where?, and why? in the abstraction of fundamental questions in his framework. Dane (2010) applied the same six interrogatives for evaluation of research.

Guide to applying CT in IS research

Through IS research, CT helps to manage and regulate the ambiguities in IT solutions, including the complexities in organisation. In the IS domain, CT has been applied to examine the success of IS planning, assessment, and evaluation. Although CT is increasingly being applied in IS research, pitfalls remain, primarily because of the lack of a guide, which limits individuals' ability at both micro and macro levels. Strategy is core to organisations' vision for improved performance, competitiveness, and sustainability. It focuses mainly on technology, people, and process in the context of IS. It directs the functions of the organisation in the long term. It allows the organisation to be sustainable towards its competitors. Performance is evaluated according to organisational performance in a period of time, to measure whether the organisation is successful or not.

There is no one universal method or best way to manage that can be used in every IS and technologies problematised situation. Sarker, Sarker, and Sidorova (2006) describe problematisation as the first stage of moments of translation, during which a focal actor formulates a problem on its own terms, identifies other relevant actors, and indicates how the problem affects the other actors. Each situation requires the consideration of a wide range of external and internal factors. This approach is intended to focus on individual actions and best fits. CT, using the Blanton et al. (1992) model, was employed as a lens to analyse the data.

Table 6.1 Analysis components

Component	*Guide*
What	Provides insights that are useful in identifying actors and factors in the deployment of IT solutions. Also, it helps to identify the attributes that form the basis of each IT solution, which ensures consistency, focus, and quality.
Who	This helps to identify the actors explicitly or implicitly involved in the activities and processes, whether in the development, implementation, management, or governance of IT solutions in an organisation. By identifying the actors, it helps to ensure balance of participants and units, and align them with performance. Also, this assists to understand the roles of individuals and groups in the process of carrying out the activities of IT solutions in an environment.
Why	This helps to identify the rationalities for employing or not employing each IT solution. It helps to draw on and enforce governance on the IT solutions. In addition, comprehension contributes to the performance of an organisation.
How	This guides the suitability of methodology (methods, approaches, and techniques) that are applied in the deployment, management, governance, or use of IT solutions in an organisation. It includes complementarity of methodologies, including the use of CT for data analysis or interpretation.
When	The activities and processes involved in the deployment, governance, or use of IT solutions are guided by periods of events within an environment. This is to ensure context and relevance. Thus, it ensures assessments to detect deficiency or obsolescence in various IT solutions. Also, it enables creating criteria for the evaluation of the life-span of IT solutions.
Where	Each IT solution has its limit and capability; hence scope is always critical to maintain consistency and uniformity in attaining efficiency and effectiveness. The scope ensures that there are no omissions of any particular areas or IT solutions in enabling and supporting organisations for performance improvement.

As explained above, CT consists of structure, strategy, environment, and performance. The theory and Tables 6.1–6.4 can complementarily guide the analysis of the data:

The application of CT to a study enables the researcher to examine and recognise that the successful application of a technique or method in one situation or environment does not guarantee success in another. Instead, the researcher will be able to examine and understand each case in terms of how it is affected or influenced by factors such as contextual, organisational, environmental, and human dimensions. As shown in the Tables 6.2–6.5, the CT tenets and categories shape and put things in perspective and through which responses are obtained in context, using the component tools. An actor is considered a common subject or entity in IS studies and industry. An actor is defined as anything that can or has the potential to make a difference in an environment (Callon, 1986). Actors are involved in all aspects of IS and technologies deployment that constitutes development and implementation.

An organised, programmatic, and coordinated process of actions should be followed in assessing and evaluating IT solutions in an environment. An environmental contingency is important to assess the effort required to transform and change an environment. The effects of transformation or environmental changes are determined from Table 6.2 results. This is because it is not possible to accurately predict possible changes that the deployment and practices of IT solutions bring to the environment.

There are two primary underlying assumptions: (1) there is no one best way or solution to structure an environment (Joslin & Müller, 2016); and (2) all structures are not equally effective under the same circumstances. By implication, internal systems, structures, and processes should 'fit' with external environment factors in the deployment, support, maintenance, and use of IT solutions in an organisation. Thus, Table 6.3 helps to extricate information and factors that can enable efficiency and effectiveness of structure, which in essence depends on the degree of fit.

Strategy focuses on tasks that advance the interest of an organisation through enabling tools, such as IT solutions. The theory helps to objectify technological techniques that conform to solving organisations' challenges. Table 6.4 provides an approach for assessing and managing capability. Through the table, interconnection and relationship between systems are structurally collected for strategic purposes. Evaluation of IT solutions facilitates the formulation of strategy in an organisation. Such evaluation processes are to promote communication and provide usefulness. Strategy facilitates better relationships and the improvement of IT solutions in organisations. From a CT perspective, strategy is fundamental and an important factor in managing IT solutions.

Performance is a tool for organisational sustainability and competitiveness. Thus, performance can be overwhelmingly difficult to assess, primarily because it is associated with various internal and external factors

Table 6.2 CT: environmental

Tenet	*Category*	*Component*					
		What	*Who*	*Why*	*How*	*When*	*Where*
Environment	Technology	What IT solutions are deployed and used? What IT solutions are required and used?	Who are the actors and what are their roles in the deployment of IT solutions in the organisation?	Why are certain IT solutions deployed and used as opposed to others?	How are the IT solutions deployed and used?	When are certain IT solutions deployed and used in an organisation?	Where are the IT solutions deployed and used in an organisation?
	Process	What are the processes that are followed in the deployment of IT solutions in the environment?	Who are the actors that are responsible for the formation and governance of processes in the deployment of IT solutions?	Why are certain processes employed in the deployment of IT solutions?	How are certain processes used as guidelines in the deployment of IT solutions?	When are processes employed in the deployment and use of IT solutions?	Where are processes employed in the deployment and use of IT solutions?
	People	What are the roles of the people that are involved in the deployment of IT solutions in the environment?	Which actors are responsible for different tasks in the deployment and use of IT solutions in the organisation?	Why are certain people involved in the deployment of the different IT solutions?	How are individuals and groups involved in the deployment of IT solutions in the organisation?	When are individuals and groups involved in the deployment of IT solutions in an organisation?	Where are individuals and groups involved in the deployment of IT solutions in an organisation?

Table 6.3 CT: structure

Tenet	*Category*	*Component*					
		What	*Who*	*Why*	*How*	*When*	*Where*
Structure	Technology	What are the IT solutions that are deployed and used to enable and support governance and management of the organisational structure?	Who are the actors that are responsible for the deployment of IT solutions for the governance and management of the organisational structure?	Why are certain IT solutions employed in the structure of the organisation?	How do IT solutions influence the organisational structure, and vice versa?	When are different IT solutions deployed and used for the governance and management (e.g., accessibility and system control) of the organisational structure?	Where are different IT solutions deployed and used for the governance and management (e.g., accessibility and system control) of the organisational structure?
	Process	What IT solutions' processes are deployed and used for the governance and management of the organisational structure?	Who are the actors that are responsible for the formation and management of IT solutions within the organisational structure?	Why are the different processes employed in the structure of the organisation?	How are the different IT solutions' processes deployed and used for the governance and management of the organisational structure?	When are the different IT solutions' processes deployed and used for the governance and management of the organisational structure?	Where are the different IT solutions' processes deployed and used for the governance and management of the organisational structure?
	People	What are the roles of the people in the organisational structure that are responsible for the deployment of IT solutions?	Who are the actors that are responsible for the deployment, governance, and management of IT solutions within the organisational structure?	Why are the different roles defined in the structure in the manner that they are?	How are the various actors involved in the deployment, governance, and management of IT solutions within the organisational structure?	When are the various actors involved in the deployment, governance, and management of IT solutions within the organisational structure?	Where are the various actors involved in the deployment, governance, and management of IT solutions within the organisational structure?

Table 6.4 CT: strategy

Tenet	*Category*	*Component*					
		What	*Who*	*Why*	*How*	*When*	*Where*
Strategy	Technology	What IT solutions form the core of strategy in the organisation?	Who are the actors responsible for the formation and governance of IT solutions for the organisational strategy?	Why are certain IT solutions defined in the organisational strategy in the manner that they are?	How are certain IT solutions defined (selected, developed, or implemented) in the enablement of organisational strategy?	When are certain IT solutions defined (selected, developed, or implemented) in the organisational strategy for the governance and management of activities?	Where are the IT solutions defined in the organisational strategy?
	Process	What processes are followed in the development and implementation of the IT strategy in the organisation?	Who are the actors responsible for the formation, processes, and governance of the IT strategy in the organisation?	Why are certain processes employed in the development and implementation of the IT strategy?	How are processes used to develop and implement the IT strategy?	When (levels or stages) are processes used in the development or implementation of the IT strategy in the organisation?	Where (units) are processes used in the development or implementation of the IT strategy in the organisation?
	People	What roles are involved in the development or implementation of the IT strategy in the organisation?	Who are the actors that are responsible for the formation and management of the IT strategy in the organisation?	Why are certain people involved in the development or implementation of the IT strategy in the organisation?	How are certain people involved in the development or implementation of the IT strategy in the organisation?	When are certain people involved in the development or implementation of the IT strategy in the organisation?	Where (units) are people involved in the development or implementation of strategy in the organisation?

and characteristics. Increasingly, IT solutions are employed to achieve performance. Hence, it is essential and critical to continually assess and evaluate IT solutions to ensure they fit the vision and mission, including the requirements of an organisation. Through Table 6.5, the relationship between components of IT solutions can be researched, traced, holistically assessed, evaluated, and monitored, in enabling and supporting an organisation towards achieving a desired state of performance. In essence, CT proposes that performance outcomes of an organisational unit are a result of the fit between the unit's external context and internal arrangements.

Tables 6.2–6.5 answer some fundamental questions that drive IS research, such as the following: How can the focal actors involved in both IT solutions and organisational activities be determined in an environment? What are the critical factors? How can the influence of critical factors on an environment be exhumed from the perspective of IT solutions' deployment and use? Tables 6.2–6.5 also help to bridge the gap, and ensure alignment between IT and business units, in that they holistically cover IT solutions' activities in the organisational contexts of environment, structure, strategy, and performance.

A structure fits routine technological alignment, people's skill set, and process centredness of an organic (participatory) nature. In summary, there is no single formula or entity that best fits all organisations' needs and patterns. Attempts to answer questions by the different components, from various categories, in the areas of the CT tenets help to find unique fits for individual organisations. As a result, organisations will not be equally efficient. Through the use of this guide in studies, researchers can, in the deeper sense of things, suggest and recommend necessary solutions which can help organisations address their unique challenges.

Conclusion

CT is theoretical, practical, and methodological, as reaffirmed in this chapter. This has been well documented. However, the theoretical nature provides *food for thought*. It helps to follow and understand how and why things fit together philosophically and conceptually. Even though it is difficult to apply in reality and in natural settings, it does help to broaden thinking. In its practical use, as shown in this chapter, it is equally important for every organisation, in that its application brings a fresh perspective that enables both academics and professionals, including managers, to examine the enabling and constraining factors separately in each case. Thus, it brings numerous possible solutions to a problem. Hence, its application in IS studies is vital. Tables 6.2–6.5 express the methodology nature of CT, which can be applicable to both academics and businesses. The methodology can be employed to guide analysis of data in IS studies. It can also be employed by organisations to measure and validate their processes and activities to gain a better understanding of why things happen in the manner that they do.

Table 6.5 CT: performance

Tenet	*Category*	*Component*					
		What	*Who*	*Why*	*How*	*When*	*Where*
Performance	Technology	What IT solutions are deployed and used specifically for performance improvement in the organisation?	Who are the actors specifically tasked to ensure that the use of IT solutions improves the performance of the organisation?	Why are certain IT solutions deployed, specifically for organisational performance?	How are different IT solutions deployed and used, specifically for performance improvement in the organisation?	When are different IT solutions deployed and used, specifically for performance improvement in the organisation?	Where are certain IT solutions deployed and used, specifically for performance improvement in the organisation?
	Process	What IT solutions' processes are followed specifically for performance improvement in the organisation?	Who are the actors responsible for the formation and governance of processes in the deployment of IT solutions for organisational performance?	Why are different IT solutions' processes employed for organisational performance?	How are different IT solutions' processes enforced, specifically for performance improvement in the organisation?	When are different IT solutions' processes followed, specifically for performance improvement in the organisation?	Where are certain IT solutions' processes followed, specifically for performance improvement in the organisation?
	People	What roles are involved in the improvement (evaluation and assessment) of performance in the organisation?	Who are the actors that ensure IT solutions' management for organisational performance?	Why are different people tasked with certain IT solutions, purposely to improve performance in the organisation?	How are different people tasked with IT solutions, purposely for performance improvement in the organisation?	When are different people tasked with certain IT solutions for the performance of the organisation?	Where are people tasked with IT solutions, specifically to improve performance in the organisation?

However, it has never been easy to apply the theory in IS research. I do hope that the guide presented in this chapter will help students and other researchers in their zeal to contribute to the business world through their studies. It is also more difficult for many managers who have interest in the application of CT but do not have a strong background of academic rigour. This chapter is intended to be of utmost guidance to enable them to improve workers' satisfaction. Employees will therefore be in a better position to understand the fit within the organisational structure. Through the model, CT provides a methodological approach towards the organisational structure that leads to a better fit in the overall organisation with resultant effective performance.

References

Betts, S. C. (2003). Contingency theory: Science or technology? *Journal of Business & Economics Research (JBER)*, *1*(8), 123–130.

Blanton, J. E., Watson, H. J., & Moody, J. (1992). Toward a better understanding of information technology organization: A comparative case study. *MIS Quarterly*, *16*(4), 531–555.

Callon, M. (1986). Some elements of the sociology of translation: Domestication of the scallops and the fisherman of St Brieuc Bay. In: J. Law (ed.), *A new sociology of knowledge, power, action and belief*, pp. 196–233. London: Routledge.

Creswell, J.W. (2003). *Research design: Qualitative, quantitative, and mixed methods approaches* (2nd ed.). Thousand Oaks, CA: Sage.

Dane, F. C. (2010). *Evaluating research: Methodology for people who need to read research*. London: Sage.

Davenport, T. H., & Short, J. E. (2003). Information technology and business process redesign. *Operations Management: Critical Perspectives on Business and Management*, *1*(97), 97–123.

Denzin, N. K., and Lincoln, Y. S. (2000). Strategies of inquiry. *Handbook of qualitative research* (2nd ed.), pp. 367–378. London: Sage

Drazin, R., & Van de Ven, A. H. (1985). Alternative forms of fit in contingency theory. *Administrative Science Quarterly*, *30*(4), 514–539.

Donaldson, L. (2001). *The contingency theory of organisations*. London: SAGE.

Donaldson, L. (2006). The contingency theory of organisational design: Challenges and opportunities. In *Organisation design* (Vol. 6), pp. 19–40. Boston, MA: Springer.

Fiedler, F. E. (1964). A contingency model of leadership effectiveness. *Journal for Advances in Experimental Social Psychology*, *1*(12), 149–190.

Giddens, A. (1984). *The constitution of society: Outline of the theory of structuration*. Cambridge: John Polity Press.

Hackney, R., Burn, J., & Dhillon, G. (2000). Challenging assumptions for strategic information systems planning: Theoretical perspectives. *Communications of the Association for Information Systems*, *3*(9), 2–23.

Holliday, A. (2007). *Doing & writing qualitative research*. Newbury Park: Sage.

Howell, D., Seidel, R., & Windahl, C. (2010). A project contingency framework based on uncertainty and its consequences. *The International Journal of Project Management*, *28*, 256–264.

Husted, B. W. (2000). Contingency theory of corporate social performance. *Business & Society*, *39*(1), 24–48.

Iyamu, T., & Adelakun, O. (2008). The impact of non-technical factors on information technology strategy and e-business. *In the Proceedings of Pacific PACIS 2008, 281.*

Joslin, R., & Müller, R. (2016). The impact of project methodologies on project success in different project environments. *International Journal of Managing Projects in Business*, *9*(2), 364–388.

Maruping, L. M., Venkatesh, V., Thong, J. Y., & Zhang, X. (2019). A risk mitigation framework for information technology projects: A cultural contingency perspective. *Journal of Management Information Systems*, *36*(1), 120–157.

Mertens, D.M. (2005). *Research methods in education and psychology: Integrating diversity with quantitative and qualitative approaches* (2nd ed.). Thousand Oaks, CA: Sage.

Mooney, G. H., Gurbaxani, V., & Kraemer, L. K. (1996). A process oriented framework for assessing the business value of information technology. *ACM SIGMIS Database*, *27*(2), 68–81.

Orlikowski, W. (1992). The duality of technology: Rethinking the concept of technology in organisations. *Organisational Sciences*, *3*(3), 398–427.

Sarker, S., Sarker, S., & Sidorova, A. (2006). Understanding business process change failure: An actor-network perspective. *Journal of Management Information Systems*, *23*(1), 51–86.

Sekaran, U. (2003). *Research methods for business* (4th ed.). New York: John Wiley & Sons, Inc.

Stoel, M. D., & Muhanna, W. A. (2009). IT capabilities and firm performance: A contingency analysis of the role of industry and IT capability type. *Information & Management*, *46*(3), 181–189.

Van de Ven, A. H., Ganco, M., & Hinings, C. R. (2013). Returning to the frontier of contingency theory of organizational and institutional designs. *Academy of Management Annals*, *7*(1), 393–440.

Van Looy, A., & Van den Bergh, J. (2018). The effect of organization size and sector on adopting business process management. *Business & Information Systems Engineering*, *60*(6), 479–491.

Zachman, J. A. (1987). A framework for information systems architecture. *IBM Systems Journal*, *26*(3), 276–292.

7 Diffusion of innovation

Table of contents

Introduction

In recent years, the roles of technological innovation on the economy and knowledge-based activities have increased rapidly. Along the same line, there is a substantial set of evidence from studies to suggest that innovation, especially the technological aspect, constitutes a crucial factor for economic growth. Martins and Tarblanche (2003) argued that innovation is the implementation of a new and possible problem-solving idea and practice regarded as new by the relevant unit of adoption and through which change is brought about.

In the technology space, innovation can be split into two main aspects. First, some people, including organisations, create new solutions. Second, others create various new ways of using the solutions to achieve or arrive at the same objectives. Many organisations focus on the utilisation of technology, and others are more interested in innovations for their goals and objectives. Some individuals and organisations take it further by producing and reproducing technology over time and space through innovations. What is even more important and challenging is illuminating the practices

DOI: 10.4324/9781003184119-7

and philosophical assumptions of the symmetry between technology innovation and the social context, which diffusion of innovation (DOI) helps to examine.

Innovation can manifest as an end to a means in practice. Some studies argued that innovation is not a single act, but rather a process which begins with an idea or an invention (Damanpour, 1991; Dougherty & Bowman, 1995; Drucker, 2002; Tidd & Bessant, 2018). According to Corso and Pavesi (2000), innovation is a continuous and cross-functional process of learning and improvement which involves the integration of the current with the future.

The definitions and exploits provided by organisations and studies make it difficult to diffuse systems, technologies, and innovations within environments. DOI makes a case in this regard, through its lens, to examine why things happen in the way that they do in the computing environment. The DOI theory focuses on the manner in which a new technological idea, artefact, or technique, or a new use of an old one, migrates from creation to use and management. According to Rogers (2003), in DOI theory, technological innovation is communicated through particular channels, over time, among the members of a social system.

Although organisations have long realised that the approval or rejection of information technology (IT) solutions is a prerequisite for its success or failure, it is still a widespread problem. It is for this reason that users' approval of new technology, based on the ability to diffuse it, has become such an integral part of information systems (IS) research and practice (Venkatesh & Davis, 2000). User approval has long been a barrier to the success of IT solutions, but if approval is achieved, so is the objective to enhance work-related performance, which is the primary aim of many organisations' IS (Davis, 1993). Jonash and Somerlatte (1999) argue that the way IS are seen and used in many organisations is different.

Based on the challenges that occur in the process of diffusing systems, technologies, and innovations within environments, IS studies are increasingly applying DOI. However, there remain challenges associated with how DOI is applied in IS research. The remainder of this chapter is divided into four main sections. The first section covers an overview of DOI, including elements of the DOI and perceived attributes of innovations. In the second section, IS and DOI are discussed. The third and fourth sections present IS research and DOI, and the use of the theory in IS research, respectively. The section that follows presents a guide on how to apply DOI in IS research. The implications of applying DOI in IS research are discussed in the sixth section. Finally, the chapter is concluded.

Diffusion of innovation

DOI is concerned with the characteristics of innovation, how it embeds with decisions, from the initial stage when the knowledge is conceived to a stage of confirmation of the innovation adoption (Weigel et al., 2014). It is a theory that seeks to explain how, why, and at what rate new ideas such as technology disseminate across a community or society. In DOI, an innovation is an idea that has not been tested, or a perceived new practice or object by an individual or group, while diffusion is the process through which an innovation is communicated to a community over time (Rogers, 2003). Although DOI is broadly about an idea, or practice or object, the concept is confined to IT solutions and IS research in the context of this chapter and book.

Elements of the DOI

There are five main elements of DOI: innovation, communication channels, time, and the social system (Rogers, 2003). It is thus a process by which (1) an innovation is (2) adopted (3) communicated through certain (4) channels over time, and (5) among the members of social system.

i *Innovation* – Broadly, innovation is relative to context and relevance as any idea, practice, or object that is perceived as new by an individual, group, or community. Rogers (2003) emphatically explains that, in an innovation, newness does not necessarily only refer to the new knowledge, but it might also refer to knowledge known about an innovation for some time but not yet developed. In addition, innovation can be something that someone had a favourable or unfavourable attitude towards but had not adopted or rejected. Innovation 'Newness' can be expressed in terms of knowledge, persuasion, or the decision to adopt (Rogers, 2003: 12). Newness qualifies an innovation for study or further investigation.

ii *Adopters* – Adopters are the minimal unit of analysis. In most studies, adopters are individuals, but can also be organisations, clusters within social networks, or countries (Meyer, 2004). Rogers (1995) defines an adopter as a classification of individuals within a social system, based on innovativeness standardised over a length of time.

iii *Communications Channels* – Communication channels have the capacity to and allow the transfer of information from one unit to another (17). Rogers (2003) defines communication as the process by which participants create and share information with one another in order to reach a mutual understanding. The essence of the diffusion process is the information exchange through which one individual communicates a new idea to one or several others, and, at its most elementary form, the process involves: (1) an innovation; (2) an individual or other unit of adoption that has knowledge of, or experience with, the innovation; and (3) a communication channel. A communication channel is the means by which messages get from one individual to another (2003: 18).

iv *Time* – Time is the third element of the diffusion process. It is necessary in terms of adoption of innovation and is rarely on an instantaneous basis. Rogers (2003: 20) states that 'much other behavioural science research is timeless in the sense that the time dimension is simply ignored or does not matter'. However, he specifies that the time dimension is involved in: (1) the innovation-decision process by which some individuals pass from first knowledge of an innovation through its adoption (i.e., the relative earliness/lateness with which an innovation is adopted) as compared with other members of the system; and (2) an innovation's rate of adoption in a system, usually measured as the number of members of the system who adopt the innovation in a given time period.

v *Social System* – The social system is the combination of external and internal influences. Rogers (2003: 23) defines social system as a 'set of interrelated units that are engaged in joint problem solving to accomplish a common goal'. However, he mentions that diffusion occurs within a social system and the social structure affects the innovation's diffusion in several ways. The social system provides a medium through which, and boundaries within which, innovation is adopted. There are many roles in a social system and their combination represents the total influences on a potential adopter.

Perceived attributes of innovation

The perceived attributes of innovation from the perspective of DOI are five: relative advantage, compatibility, complexity, trialability, and observability. From the attributes, it should not be assumed, as it sometimes has been in the past, that all innovations are equivalent units of analysis. 'This assumption is a gross oversimplification' (Rogers, 2003: 15). The following are characteristics of innovation, as can be perceived by individuals or groups, to help explain why and how innovations are adopted or rejected at the rates at which they are, according to Rogers (2003:15).

i *Relative advantage* – it is the degree to which an innovation (such as an IT solution) is viewed as superior or better than the existing idea or solution it replaces.

ii *Compatibility* – it is comparing a situation to determine the consistency of an innovation in terms of values, experiences, and needs of the potential adopters.

iii *Complexity* – it is the extent to which an innovation is difficult to either understand or use in an environment.

iv *Trialability* – it is the extent to which an innovation can be tested or experimented before a pledge to adopt it is made. In other words, new ideas that have been tried may be adopted quicker than the ones that have not been proven.

v *Observability* – it is the extent to which the innovation provides tangible results. This means the adoption of an innovation is more likely when individuals see or experience the results, in that visibility stimulates peer discussion of new ideas.

Initially, Rogers (1995) stated that diffusion occurs through a five-step decision-making process. Rogers's five stages (steps) of awareness, interest, evaluation, trial, and adoption are integral to this theory. As time progressed, Rogers (2003) changed the terminology of the five stages to knowledge, persuasion, decision, implementation, and confirmation.

IS and DOI

IT solutions innovation begins with the organisational environment, whereby there is top-down support for project management practices and investment by the organisation in the innovation process. Swanson (1994) further argues that there are three types of IS innovations. Type 1 innovations are confined to the IS task; type 2 innovations support business and administration functions; and type 3 innovations are those that are embedded in the core technology of the business. Each of these types of innovations is critical to the success of an organisation and therefore requires attention during system implementation. This leads to the discussion on IS and DOI.

IT solutions

IT solutions are used for different purposes and in various ways. The technology solutions are used in delivering information, including business processes and logics, towards achieving the goals and objectives of an organisation. Long ago, Kendall (1992) advanced the view that an IS exists only to serve the business systems of which it is a component. Some of the IT-based solutions are used to collect, store, process information, including analysis (or analytics), and report data from various sources to help with improved managerial decision-making in organisations. In addition, through IT solutions, some organisations execute their business strategies in attempts to realise the ultimate goals of competitiveness and growth. Thus, many organisations make use of IT solutions as tools to facilitate and foster various innovations in order to support and enable processes and activities.

Due to the criticality of IT solutions (IS and technologies), their development, implementation, and practice, organisations, including academic studies, follow certain norms over time and space in achieving its purposes. This requires a diffusion of the IT solutions in order to understand why things happen in the way that they do at the different stages of development, implementation, and practice of the solutions. In an attempt to make sense of IT solutions from the DOI perspective, Rogers (2003) suggested that the norm should be divided into distributed segments of five categories for individual and organisational innovativeness.

Diffusion of innovation

DOI is a process by which an idea, practice, or object that is perceived as new by an individual or other unit of adoption is communicated through certain channels over time among the members of a social system (Rogers, 2003). The theory describes the social process of communicating a new idea among the members of a community over time. The focus of the theory is not only on awareness and knowledge, but also on attitudinal change and the decision-making process that lead to the practice or adoption of an innovation (Rogers & Singhal, 1996).

The process of diffusion involves communication between change agents to encourage the adoption of an innovation by an individual, organisation, or other adoption unit through social systems. Diffusion also includes the awareness and knowledge about an innovation, and focuses on examining attitude towards change, decision-making, and the implementation of an innovation in an environment (Sutherland, 2003). One of the most difficult challenges is to identify and implement the correct or appropriate change (Oliver, 1996). The interaction between IS studies and the use of DOI is attributed to and shaped by two main factors: the type of data that is collected and how DOI is employed in the analysis of the data.

IS research and DOI

Although research methods and data collection techniques, including data sources, are extensively covered in Chapters 2 and 3, they are again briefly discussed here, purposely to give context to Figure 7.1. The outcome of a research is primarily influenced and shaped by the type of data that was collected and how the analysis of the data was carried out. The type of data, whether qualitative or quantitative, or both, determines and influences how and when technologies or innovations are diffused within an environment.

The DOI of systems and technologies in an environment is influenced and shaped by interaction amongst actors in a social context. The interaction requires an interpretive approach to make sense of the phenomenon under study. Hence, it is suitable to employ DOI for the analysis or interpretation of qualitative data in IS studies. According to Walsham (1993, 1995), many IS studies using a qualitative approach adopt an interpretive stance. Qualitative research methods are designed to help researchers understand people and the social and cultural contexts within which they live. Kaplan and Maxwell (2005) argued that the goal of understanding a phenomenon from the point of view of the participants and its particular social and institutional context is largely lost when textual data are quantified, instead of qualified.

The qualitative research method consists of different techniques which are primarily interviews, documentation, and observation. Qualitative research studies often focus on and are used to understand and explain social phenomena. There has been a general shift in IS research away from technological to managerial and organisational issues, hence an increasing interest in the application of qualitative research methods (Myers, 1997).

In order to be able to examine and understand the interaction that happens amongst actors and technologies, an in-depth interpretation that is

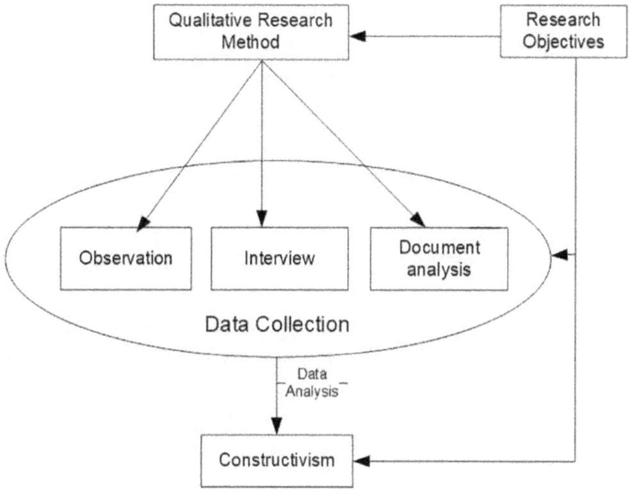

Figure 7.1 Qualitative data.
Source: Author.

guided by a framework must be drawn. As shown in Figure 7.1, the choice of qualitative method is determined and informed by the objectives of the research. Any or a combination of qualitative research techniques could be used for the collection of data. As presented in Figure 7.1, observation, interview (e.g., semi-structured), and documentation techniques are mainly used for qualitative data collection. The data are interpretively analysed to enact a constructivism conclusion to demonstrate how events are produced over time and space.

Also, in IS studies, where qualitative data are collected, this often includes 'an array of interpretive techniques which seek to describe, decode, translate, and otherwise come to terms with meaning, not frequency' (Cooper & Schindler, 2006: 196). One of the strengths of a qualitative method is that it allows relationships between agents to be intertwined, as well as makes meaning of the social context and experience, particularly from a hermeneutics perspective (Denzin & Lincoln, 2005). In Iyamu's (2010) description, agents which form a relationship and interact in a social system are both human and technology. Qualitative data can be gathered from multiple sources such as people, texts, objects, artefacts, media, settings, and environments.

i Observation technique – Observation is a way of gathering data by watching behaviour and events or noting physical characteristics in their natural setting. Observations can be overt (participants know that they are being observed) or covert (participants are not aware that they are being observed and the observer is concealed). The benefit of covert observation is that people are more likely to behave naturally if they are not aware that they are being observed. The observation technique provides the opportunity to document activities, behaviour, and physical aspects within context. The technique eradicates the dependence upon humans' willingness, interest, and ability to respond to questions. The observation technique is valuable for collecting qualitative data.

ii The interview approach – The nature of the interview (e.g., semi-structured) approach, which allows for descriptive and qualifying type of data, makes a strong case for qualitative data. According to Kvale (1996), the most useful interview format for conducting qualitative research is often 'semi-structured'. This means the interview is not highly structured, as is the case of an interview that consists of all close-ended questions, nor is it unstructured, so that the interviewee is simply given a licence to talk freely about the topic.

iii Documentation (document analysis) technique – As part of data gathering, the researcher seeks access to the case studies (organisations') documentation relating to the study. This includes items such as the organisation's strategy documents and documentation manuals on IS and reports.

iv Data analysis – Analysis of data is critical to any study. The analysis depends on the type of data that is collected. Mkhomazi and Iyamu (2013) argued that analysis of data is fundamental to any study, and that analysis of the data determines and shapes the results of the study. It is therefore important that the appropriate theory is selected to underpin the study for analysis purposes. For example, a sociotechnical theory such as DOI should be employed to analyse qualitative data, based on the objectives of the study, to ensure constructivism of the results.

DOI for analysis in IS research

The innovation-decision process is often used as a lens in IS research for the analysis or interpretation of data. The model of innovation-decision process from the perspective of DOI has been used to underpin many IS studies in recent years. The theory is employed to analyse qualitative data. As depicted in Figure 7.2, there are five components in the model, namely, knowledge, persuasion, decision, implementation, and confirmation. The components guide, in sequence, the data analysis within the context and objectives of the studies.

Brief descriptions of the component are provided as follows:

i Knowledge

This is the first stage of the decision process. It is a stage where an individual or a group is first exposed to an innovation. However, the individual or group does not have information about the innovation. As

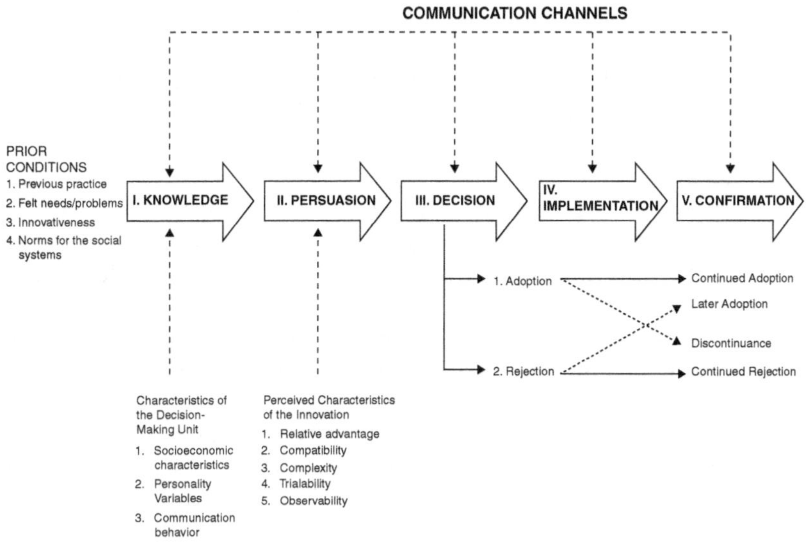

Figure 7.2 Innovation-decision process (Rogers, 2003).
Source: Author.

a result, the individual is not motivated or encouraged to find out more information about the innovation. This inspires research to investigate how an individual acquires knowledge about innovation of an IT solution and how and when the knowledge is applied. Schultze and Leidner (2002) argued that in order to understand the ways that IS can support the management of knowledge in organisations, consideration must be given to not only the intended positive consequences of knowledge and its management, but also the unintended negative ones.

The way and why IT solutions are deployed and used in organisations are different. This can be attributed to the knowledge which employees have of the innovation (IT solution). Also, the source of knowledge gathering is critical, as it shapes the understanding of the innovation, which ultimately leads to usefulness or rejection. An organisation can formulate more creative strategies in a more frequent and differentiated manner. This is aimed at creating a knowledge-based strategy for competitive advantage.

ii Persuasion

In the second stage, persuasion, an individual or group becomes interested in the innovation of an IT solution. Also, the individual actively seeks more information related to the IT solution. Thus, the persuasion component is used as a guide in research to elicit information about when individuals or decision-making units exhibit favourable or unfavourable actions towards innovation of IT solutions in an environment. At the persuasion component of IT solutions' deployments, a person's focus or thinking can be affected by, or related to, feelings or perception. Some employees in the organisation develop an attitude towards certain innovations and are psychologically involved with IT solutions, including their processes and activities.

iii Decision

In the third stage, decisions are made. It is the stage where individuals weigh the pros and cons of changing from current to another platform, using the innovation of IT solutions. Thus, each individual decides whether to adopt or reject the innovation. Due to the individualistic nature of this stage, Rogers (2003) suggests that it is the most difficult stage in which to acquire empirical evidence. The process of decision-making in the deployment and use of IT solutions is inherently vital for individuals who rely on them for survival and organisations that depend on the solutions (innovation) for competitiveness and sustainability. Thus, it is essential to enquire why, how, and what decision was made in each process or activity.

iv Implementation

Implementation means that an individual or a group employs the innovation of IT solutions to varying degrees, depending on the circumstances in each unit of the organisation. During implementation, individuals or groups determine the usefulness of the innovation and may search for more information about the solution. This is a crucial moment for many organisations to ensure that a correct decision has been made. This is because deployment of IT solutions or its innovation are driven by the quest to remain sustainable and competitive to avoid the collapse or liquidation of the organisation. These are the fundamental factors that have the ability and substantial value to support the decision-making process.

v Confirmation

This is the last stage, the state of finalisation of a process of innovation (Rogers, 1995). It means the stage where individuals or groups finalise the decision to continue or stop using the innovation (IT solution). At this stage, a confirmation is reached whether the individuals have made the right or appropriate decision about an IT solution on behalf of the organisation or not. Thus, the confirmation is both a cognitive dissonance and a cohesive coherence decision.

In summary, knowledge and perception indicate that the need for change has been communicated and an innovation has been identified for adoption, whereas confirmation occurs after the innovation has been evaluated and a decision has been made. As shown in Figure 7.2, Rogers' model (2003) allows an individual to advance through components of the decision-making process, based on the end result being implementation and adoption. However, Rogers demonstrates that implementation must occur in order for the innovation to be utilised.

It is on the basis of the discussion above that DOI is often applied in IS research. The explanation does not necessarily take away the challenges which researchers, particularly aspiring researchers, encounter in attempts to apply the theory. It is usually applied to assess and evaluate how IT solutions are diffused, including what decisions and at what stage they are made. This is to ascertain the consequence and impact which the decisions have on the innovation and the goals and objectives of an organisation. Thus, DOI can be applied at both micro and macro levels in the analysis or interpretation of data in IS research. This helps to elucidate information and understanding of the deployment of IT solutions for operational or strategic purposes (Venkatesh & Davis, 2000). Rogers (1995) adds that the change of strongly held attitudes usually occurs through interpersonal channels and becomes more important at the persuasion component in order to implement change within the innovation-decision process (Rogers, 1995).

Applying DOI theory in IS research

An individual or group can reject an innovation at any time, during or post the adoption process. In some organisations, such decisions are either trivialised or ignored. As a result, solutions are not found and the circumstances repeat themselves. DOI helps to examine the process by asking critical questions such as those presented in Table 7.1. The guide is intended to be a comprehensive effort for conducting research in IS to assess or evaluate the process of diffusing IT solutions with an organisation or environments.

Applying DOI in IS research is significant and critical mainly because it guides and influences the outcome of the research. However, the theory has never been straightforward or easy in its applicability. It constitutes sociotechnical interaction between humans and technical artefacts within context. The process of diffusion involves communication amongst change agents in order to encourage the adoption of an innovation by individuals and organisations through social systems. Diffusion also includes the awareness and knowledge of an innovation, and focuses on examining attitudes towards change, decision-making, and implementation of the innovation (Sutherland, 2003: 6). The real challenge is to identify and implement the correct change (Oliver, 1996: 7).

One of the primary strengths of the DOI theory is its use in the interpretation of qualitative data. The innovation-decision process of DOI consists of five components: (1) knowledge; (2) persuasion; (3) decision; (4) implementation; and (5) confirmation. The components are characterised as a process that occurs while individuals and groups participate in a series of actions related to decisions (Rogers, 1995: 162). Table 7.1 presents a guide on how to apply DOI for either analysis or interpretation of data in IS studies, focusing on technology innovation within the social context. Analysis or interpretation is about making sense from answers to questions which are relevant, within context, and meaningful to the phenomenon under study.

Table 7.1 Guide to analysis applying DOI

Component	*Who*	*How*	*Where*	*When*	*Why*
Knowledge	Who has the knowledge to undertake tasks in the innovation of IT solutions?	How was knowledge pertaining to innovation of an IT solution acquired?	Where was the relevant knowledge relating to innovation of an IT solution acquired?	When (at what stage) was the knowledge pertaining to innovation of an IT solution acquired?	Why was the knowledge pertaining to innovation of an IT solution acquired?
Persuasion	Who has the ability to persuade other actors to enrol and participate in the activities of innovating IT solutions?	How were the actors persuaded to enrol and participate in the processes and activities of innovation of IT solutions?	Where (from what stage) in the organisation were actors persuaded to partake in innovation of an IT solution?	When (at what stage) were the actors persuaded to enrol and participate in the processes and activities of innovation of an IT solution?	Why were actors persuaded to enrol and participate in the activities and processes of innovation of IT solutions in an organisation?
Decision	Who made the decisions to engage participants in the activities and processes of innovation of IT solutions?	How were the decisions relating to the innovation of IT solutions made in an organisation?	Where (from what stage) were critical decisions relating to innovation of IT solutions made?	When (at what stage) were critical decisions relating to the innovation of IT solutions made?	Why (rationale) were critical decisions made by the actors during innovation of IT solutions?
Implementation	Who were the actors that were involved in the implementation of the IT solutions' innovation?	How was the implementation of IT innovation carried out in the organisation?	Where did the actual implementation of IT solutions innovation take place (unit)?	When exactly was the implementation of IT solutions carried out in the organisation?	Why was the implementation of IT solutions innovation carried out in the way that they were?
Confirmation	Who confirmed the innovations of IT solutions, including the processes and activities in the organisation?	How were the activities and processes that were involved in the innovation of IT solutions, including the processes and activities, confirmed?	Where (unit) were the innovations of IT solutions confirmed in the organisation?	When (at what stage) were innovations of IT solutions, including the processes and activities, confirmed?	Why were innovations of IT solutions, including the processes and activities, confirmed in the organisation?

Table 7.1 provides a guide to a holistic analysis of qualitative data. The questions, as contained in the table, can be used as a checklist during analysis. What is even more important is that the questions can flexibly fit into any IS studies in achieving the research objectives.

Implication of DOI in IS research

Research studies frequently do not capture the full range of consequences of the innovations that are under investigation (Rogers, 2003). In the past, most studies pertaining to systems and technologies innovation made use of quantitative methodologies and focused wholly on outcomes that were demonstrated at a single point in time (Meyer, 2004). Sooner than later, practitioners began to realise that they were likely to encounter a wide range of consequences for the innovations they adopted, making the type of data as well as the analysis or interpretation of the data difficult.

The act of adoption of innovation and the process of DOI is shaped and influenced by actors' interactions within a social context or system. Hence in-depth interrogation of analysis as presented in Table 7.1 is required to reveal the implications and consequences of why and how things are the way that they are, in the course of innovation and diffusion of IT solutions in organisations. Research in IS takes cognisance of and acknowledges the fact that weaknesses and challenges do exist in selecting technique for data analysis or interpretation. For example, it is not easy to keep sets of qualitative data focused, primarily because of the depth of the study, including the descriptive and explanatory nature of data. However, the DOI theory allows the combination of interviews of participants from different cases (organisations). Thus, the data gathered during the study can then be cross-checked against each other for purposes of generalisation.

It is often difficult to generalise results when one case is used in a study. According to Hofstee (2006), there is difficulty in gathering data that could be used in obtaining results for generalisations. DOI helps to address this through its strength of generalised theory. In the use of multiple cases (organisations), the DOI allows a generalised theory to be developed due to the richness and depth of the data from multiple cases.

Conclusion

The contribution of this chapter is primarily on the guide that it provides on how to make use of the DOI theory for data analysis or interpretation in IS research. The chapter emphasises the criticality of the five components of the innovation-decision process in shaping the results of the study that employs it. The use of the DOI theory has impact, from both theoretical and practical perspectives, in the innovation and diffusion of systems and technologies, as presented in Table 7.1.

Another contribution of DOI in IS studies is understanding the relationship amongst the phases of the innovation-decision process, as they impact the activities and processes of organisations. This is critical to the managers of IS and technologies in gaining deeper clarity on how culture, values, and strategies fit and align with organisations' vision and mission. Also, this helps organisations to formulate relevant objectives. Understanding the relationship that flows in the innovation-decision process implies that managers can unveil influencing factors which they would not have been able to do under normal circumstances.

References

Cooper, D., & Schindler, P. (2006). *Business research methods.* New York: McGraw-Hill/ Irwin Series.

Corso, M., & Pavesi, S. (2000). How management can foster continuous product innovation. *Integrated Manufacturing Systems, 11*(3), 199–211.

Damanpour, F. (1991). Organizational innovation: A meta-analysis of effects of determinants and moderators. *Academy of Management Journal, 34*(3), 555–590.

Davis, F. D. (1993). User acceptance of information technology: System characteristics, user perceptions and behavioural impacts. *International Journal of Man-Machine Studies, 38*(3), 475–487.

Denzin, N. K., & Lincoln, Y. S. (2005). The seventh moment: Out of the past. In N. K. Denzin & Y. S. Lincoln (eds.), The landscape of qualitative research (4th ed.), pp. 611–640. Thousand Oaks, CA: Sage.

Dougherty, D., & Bowman, E. H. (1995). The effects of organizational downsizing on product innovation. *California Management Review, 37*(4), 28–44.

Drucker, P. F. (2002). The discipline of innovation. *Harvard Business Review, 80*(8), 77–83.

Hofstee, E. (2006). Constructing a Good Dissertation: A Practical Guide to Finishing a Master's, MBA or PhD on schedule, EPE Publishers, Sandton, South Africa.

Iyamu, T. (2010). Human interaction with structure in the computing environment. *Proceeding of the* 16th *Americas Conference on Information Systems, Lima, Peru, August 12–15. AIS Publication.*

Jonash, R.S., & Sommerlate, T. (1999). *The innovation premium.* Reading, MA: Perseus Books, 115–135.

Kaplan, B., & Maxwell, J. A. (2005). Qualitative research methods for evaluating computer information systems. In: J.G. Anderson & C.E. Aydin (eds.), *Evaluating the organizational impact of healthcare information systems*, pp. 30–55. New York: Springer.

Kendall, P. (1992). *Introduction to systems analysis and design: A structured approach* (2nd, ed.). Dubuque, IA: Wm. C Brown Publishers.

Kvale, S. (1996). *Interviews: An introduction to qualitative research.* London: Sage.

Martins, E., & Terblanche, F. (2003). Building organisational culture that stimulates creativity and innovation. *European Journal of Innovation Management, 6*(1), 64–75.

Meyer, G. (2004). Diffusion methodology: Time to innovate? *Journal of Health Communication, 9*(S1), 59–69.

Mkhomazi, S., & Iyamu, T. (2013). A guide to selecting theory to underpin information systems studies. In: Y. Dwivedi, H. Henrksen, D. Wastell, & R. De (eds.), *Grand successes and failures in IT*, pp. 525–538. Berlin: Springer.

Myers, M. D. (1997). Qualitative research in information systems. *Management Information Systems Quarterly*, *21*(2), 241–242.

Oliver, P. (1996). The concept of change management. In: P. Oliver (ed.), *The management of educational change: A case study approach*, pp. 1–9. Brookfield, VT: Ashgate Publishing Company.

Rogers, E. M. (1995). *Diffusion of innovations*. (4th ed.). New York: Free Press.

Rogers, E. M. (2003). *Diffusion of innovations* (5th ed.). New York: Free Press.

Rogers, E. M., & Singhal, A. (1996). Diffusion of innovations. In: M. B. Salwen and D. W. Stacks (eds.), *An integrated approach to communication theory and research*, pp. 409–420. Mahwah, NJ: Erlbaum.

Schultze, U., & Leidner, D. (2002). Studying knowledge management in information systems research: Discourses and theoretical assumptions. *Management Information Systems Quarterly*, *26*(3), 213–242.

Sutherland, P. J. (2003). Administrators perceptions of diffusion of courses with world-wide features. *Journalism & Mass Communication Educator*, *57*(4), 384–395.

Swanson, E. B. (1994). Information systems innovation among organizations. *Management Science*, *40*(9), 1069–1092.

Tidd, J., & Bessant, J. (2018). Innovation management challenges: From fads to fundamentals. *International Journal of Innovation Management*, *22*(05), 1840007.

Venkatesh, V., & Davis, F. D. (2000). A theoretical extension of the technology acceptance model: Four longitudinal field studies. *Management Science*, *46*(2), 186–204.

Walsham, G. (1993). *Interpreting information systems in organisations*. Wiley: Chichester.

Walsham, G. (1995). Interpretive case studies in IS research: Nature and method. *European Journal of Information Systems* *4*(2), 74–91.

Weigel, F. K., Hazen, B. T., Cegielski, C. G., & Hall, D. J. (2014). Diffusion of innovations and the theory of planned behaviour in information systems research: A meta analysis. *Communications of the Association for Information Systems*, *34*(3), 619–636.

8 Structuration theory

Table of contents

Introduction

Significantly, humans' roles, responsibilities, interactions, and relationships are essential in the deployment (development and implementation), governance, support, and management of information technology (IT) solutions in organisations. In the same vein, these human facets constrain and complicate the same IT solutions that they are supposed to enable and support. Despite the increasing sophistication, there are never-ending challenges with IT solutions in organisations, including government institutions and agencies. In attempts to resolve some of the challenges, some organisations replace one system (IT solution) with another, perhaps a more sophisticated version, yet the challenges remain. Also, academics continue to find solutions to the numerous challenges through their trade of research, by employing different methods and approaches. Thus, structuration theory (ST), a sociotechnical theory, has been overwhelmingly applied in many studies in the last two decades.

ST is a social theory used to explain connections between situated interactions and social structures of meaning, creation, and reproduction of social systems, norms, and power (Giddens, 1984). The theory offers a subtle and comprehensive view of a social system, from which it acknowledges the interaction of meaning, standards, and values, based on dynamic relationships that exist between different facets of society. In information systems (IS) research, ST is applied in various ways, such as to examine the role

DOI: 10.4324/9781003184119-8

of human action in the deployment of infrastructure, implementation of a system (Furumo & Melcher, 2006), assessment of actors' access to big data (Iyamu & Shaanika, 2020), and integration between non-technical and technical domains (Bernardi, 2017). This is primarily because ST views technology as an embodiment of norms, rules, and regulations that human actors recursively enact through their reproductive actions (Orlikowski, 1992). Also, ST proposes a form of social analysis that goes beyond dualistic ways of thinking (Pozzebon & Pinsonneault, 2005).

Despite the increasing use of ST in IS research, there are challenges, which can be viewed from three main angles, as follows: (1) that the theory is complex; (2) abstractive; and (3) difficult to apply in empirical studies. Pozzebon and Pinsonneault (2005) proceeded to attempt to improve the application of ST in empirical studies and work by drawing on the experience in IT research. Despite such efforts, challenges remain. This could be because it is complex to understand how social structures do not have inherent stability outside of human actions, owing to being socially constructed. This chapter presents a guide that can be applied in IS research, including empirical studies. The guide provides an approach that explicitly elicits information that can be analysed at the macro or micro level to elucidate the factors that influence human actions and how those actions can create stability in the deployment, management, governance, and use of IT solutions in organisations.

ST also focuses on structure and agency, with particular interest in the relationship between technical and non-technical factors. Giddens (1984) tries to summarise structuration as the structuring of social relations across time and space, by virtue of the duality of structure (DOS). ST regards 'structure' as rules and resources recursively implicated in social reproduction. Institutionalised features of social systems have structural properties in the sense that relationships are stabilised across time and space (Giddens, 1984). According to Pozzenbon and Pinsonneault (2005), ST encompasses and details the interaction between people and the 'social institutions' which are developed over time. The basis of ST is critical to analysing the relationship between people, norms, resources, and structures, which ultimately contribute to how people relate to IT solutions.

The chapter is organised into four main sections. The first section discusses ST, including agency, structure, and DOS. The section that follows covers ST in the context of IS research. The third section presents how DOS from a structuration perspective can be applied for data analysis in IS research. Finally, a conclusion of the chapter is drawn.

Structuration theory

ST is a social process that is concerned with actions between agents and the structural feature of an organisation. The theory recognises that human agents are enabled by the structures, the very structures that can constrain the agents. The actions in structuration are the results of previous structures. According to Gao and Lyytinen (2003), structuration is the process of reproducing DOS over time and space. The DOS is linked through modalities that are the interpretive schemes, resources, and norms; furthermore, the social structure and human interaction are 'broken down' into three dimensions, which are structure, modality, and interaction.

In ST, norms are key, and they are those valued systems and morals of the organisation which individuals would find themselves inherently abiding by in order to survive. Giddens (1984) identifies the need for survival as being based on either practical consciousness or discursive consciousness as a means of exerting a certain kind of power. From the perspective of structuration, Giddens (1984) explains discursive power as being individuals' ability to act and draw on their own personal experiences when required. He describes practical consciousness as individuals' ability to communicate their actions and motivations explicitly in terms of their knowledge and ability to tap into their recollections to enable them to analyse and understand their activities.

Employees perform their day-to-day activities within a system (an organisation). The structure is a result of the action. Also, employees (agents) have the ability to express why they do what they do; how and when they carry out the actions. However, some of them may not be able to explain the source of their actions but that does not necessarily terminate reproducing such action. An agent does certain things without particularly being aware of their actions, which is referred to as practical consciousness. Many of the actions that employees carry out in organisations intertwine in their minds as knowledge gained through practical consciousness.

The key elements of ST are agency, structure, and DOS (Giddens, 1984). Agency is both technical and non-technical agents (Pozzenbon & Pinsonneault, 2005). Structures are the 'rules and resources, recursively implicated in the reproduction of social systems' (Giddens, 1984: 377). Thus, structure should not be taken for its literal English meaning. DOS is based on a synthesis of structure and agency from the perspective of human actions (1984). The structures exist only through the actions of the agents, and the agents exist within a structure, making both elements inseparable.

Agency

Agency refers to humans' ability to act. According to Giddens (1984), human agency has the 'capacity to make a difference'. Also, the loss of the capacity to make a difference is powerlessness. Hence, agency is intimately connected to power. Agency, according to Rose and Scheepers (2001), is the ability to make a difference that can be achieved by making use of power. The power is based on and enacted in resources. The authors argued that the agents in an organisation learn how to exercise their power as is expected in the social system they are in. The use of power is determined by the individual's understanding of the levels of power they possess.

From the perspective of structuration, Rose and Scheepers (2001) equate agency to power. Agents in a system create and recreate actions based on power bestowed on them, which manifests in a power relationship. Power is associated with the exploitation of resources in an organisation. Resources, they argue, are 'structured properties of social systems' produced by agents in a system. The agents in a system do not only produce resources, but they also have the ability to control the resources. The agents in an organisation have the ability to exercise some form of power in the organisation. The resources can be allocated and split according to authoritative resources which can be derived from the coordination of the agent.

In their attempt to make sense of agency, they ultimately identify two types of power: authoritative and allocative power. Power is in essence the ability and inability to exploit resources in a social system. Rose (1998) defines allocative power as the power an individual would exercise to control material products, while authoritative power is explained as the power one exerts over people through the coordination of agents.

Structure

Structure is both the means and the outcome of the practices that constitute social systems, such as an IT solution. Structure can be seen as both the medium and the outcome of human agency in conjunction and collaboration with the facilities and or resources used. In Giddens' (1984) notion, structure refers, in social analysis,

> to the structuring properties allowing the binding of time and space in social systems; the properties which make it possible for discernibly similar social practices to exist across varying spans of time and space and which lend them a 'systemic' form.
>
> p. 55

According to Giddens (1984), structure has two components: rules and resources. Rules are seen as being either normative or interpretative:

i **Normative** – Normative rules represent structures of legitimation. The social agents translate normative rules as specific rights and obligations (e.g., norms). The rights and obligations are accompanied by sanctions or rewards. The legal institutions of a society are examples of a legitimating social structure. The normative rules, such as privileging private property rights, are translated into laws directed against theft which impose sanctions upon those members of society who act contrary to the rules and or laws.

ii **Interpretative** – Interpretative rules create signification or meaningful symbolic systems that provide ways for actors to see and interpret events. Agents reflexively apply interpretative schemes and stocks of knowledge. The result is communication among the social agents. The signifying structure represents the general rules of language. The mediating dimensions are the application of syntax or grammar and semantics that give the language its specific form and meaning, as well as the knowledge and experience of the agents.

The second component of structure is resources. Giddens (1984) identifies two categories of resources which are allocative and authoritative. Allocative resources relate to material or economic resources such as materials and natural resources, and result from human domination over nature. Authoritative resources are nonmaterial resources, such as human beings, and result from the domination of some agents by others. These resources provide the means or facilities for realising specific goals or social objectives. The realisation of such goals results in the manifestation of power by those agents controlling the resource.

In ST, structure as rules and resource are applied in and constituted out of action, a generative source of social interaction. The rules Giddens (1984) refers to are the rules that make up an organisation and make up the norms and values of the organisation. Agents functioning in a system are governed by rules, whether consciously or subconsciously. There are rules in organisations which agents are required to adhere to as moral codes. Resources are the assets and possessions an agent has and brings to an organisation. These resources include money, skills, and knowledge. Rose and Scheepers (2001) narratively suggest that structure exists as a trace of agents' knowledge evident in the actions of the agent. Those with the influence to wield power are vital in the allocation of resources in organisations. Consequently, allocation of resources is key in ensuring that implementation of social systems is successfully conducted. The resources in an organisation can be stored and used to meet the organisation's requirements.

Structures in ST are posited on the interaction between agents through modalities to create the structures. ST assumes that 'all social actors are learned in respect of the knowledge they possess and apply in the production and reproduction of day-to-day social encounters' (Giddens, 1984: 22). In the social context, structure consists of various patterns or relationships that are observable in multiplicity. Social structures do not exist without human actions. Giddens' modalities (interpretive schemes, resources, and norms) are the foundation on which human behaviour is recognised (Pozzebon & Pinsonneault, 2005). Structure can be enabling and at the same time constraining and a repertoire of actions that are recursively organised. The constraints may be brought on by the actions of the agents operating within the system when agents find themselves limited by their behaviours and behaviours of others in the system through actions performed and executed by the agent.

Also, 'People' defines the structure of a social system and, in so doing, inadvertently determines future actions to be taken by an organisation. The decisions that are made by individuals or groups can either limit or enable the organisation. The people in an organisation are placed in positions where they have to perform certain duties. In addition to managing expectations, people also guard their own interests in the organisation. It gives the leeway to assess why people generally tend to hold onto certain things, using them as bargaining tools. This is sometimes interpreted as a source of power. According to Rose (1998), bargain and source of power are achieved through individuals' ability to draw on their interpretive schemes.

Duality of structure

DOS comprises three fundamental components: structure, modality, and interaction. As shown in Figure 8.1, Giddens (1984) identifies the elements of the components as signification, domination, and legitimation. Institutions are made up of the actions of the agents in the system. Their actions are materialised by the structures. The formulation of structure is made up of the features in an organisation, such as human actions, which occur over a period of time, through the use of practices that form the organisational social structure. People draw on the existing structures in their ongoing interactions to produce and reproduce structures. According to Iyamu and Roode (2012: 7), DOS focuses on the processes through which legitimating structures evolve and are reconstituted by actions.

Human action is composed of structures of meaning, power, and moral frameworks. Giddens (1984) offers three modalities that link human action to social structure, as shown in Figure 8.1. These are interpretative schemes

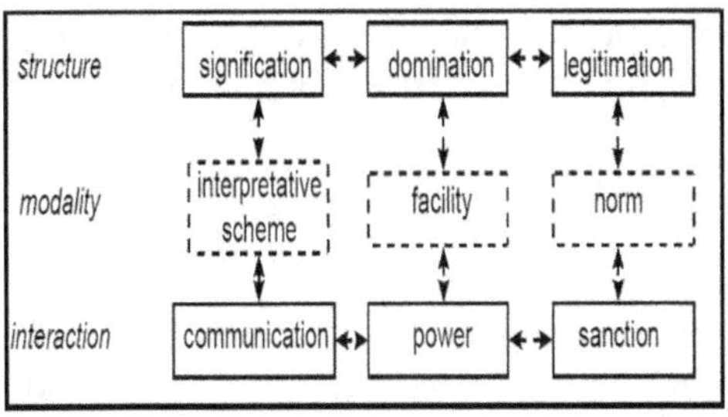

Figure 8.1 Dimensions of DOS.
Source: Giddens (1984).

(the structure of signification), resources (the structure of domination), and norms (the structure of legitimation):

- i Interpretative schemes: standardised and shared stocks of knowledge that people use to interpret behaviour and events, and thereby achieve meaningful interaction.
- ii Resources: either authoritative or allocative, they are the means for exercising power, accomplishing goals, and realising intentions.
- iii Norms: rules for sanctioned or appropriate conduct, defining the legitimacy of interaction within a setting's moral order (Orlikowski & Robey, 1991).

ST places emphasis on DOS (see Figure 8.1). According to Giddens (1984), the three modalities defined for gaining an understanding of the interaction that happens in an organisation as a social structure are interpretive schemes, facility, and norms. Interpretive schemes refer to the communication of meaning in the organisation, while resources refer to the actual people, and norms refer to the rules that govern the organisation (Pozzebon & Pinsonneault, 2005).

Structure

Structures are the type of rules and resources, including the capabilities, that are implicated in reproduction in a social system (Gao & Lyytinen, 2003). A social system comprises space and time. According to Giddens, ST is produced over repetitious actions performed by agents. Social structures are defined to exist in an organisation's traditions, norms, and values. The actions can change with time, although they are repetitious.

Structure is the understanding that structure is found in human actions, manifested in their patterns and relationships, which are affected by space and time where the social structure is informed by agents (Falkheimer, 2007). This is used interchangeably with the employees in an organisation. The tenets of DOS, which make up the structures within the theory, are signification, domination, and legitimation. According to Algesheimer and Gurău (2008), the structures are created on an individual level as well as on a group level.

Modality

The modalities (interpretive schemes, facilities, and norms) model links the process of interaction (communication, power, and sanction) with the structural components (signification, domination, and legitimation) of social systems. The modalities are also the interaction between the agents and the structures in a social system, which are closely connected, and yet the agents are capable of acting freely within the system (Giddens, 1984).

ST explains the way in which individuals' actions in a system produce structure. Social structures in an organisation are the norms, traditions, and values inherently adopted by individuals. The shared norms, values, philosophies, ideologies, assumptions, beliefs, expectations, and attitudes are some of the factors that bind individuals and groups in an organisation. The individuals and groups work together to achieve the needs and objectives of an organisation and ultimately facilitate the organisational strategy, including IT solutions. The relationship between IT solutions and agents is therefore explored and translated to gain insights into how agents interact and continuously draw from their interpretive schemes, facilities, and norms to produce signification, domination, and legitimation, respectively.

Interaction

Communication, power, and sanction are the properties of interaction; see Figure 8.2. ST explores the influence of power on resources, communication channels, and values. Through the theory, Giddens defines the duality of modality, which he explains through the DOS model; refer to Figure 8.2. The properties enable and can at the same time constrain an organisation in its quest for achieving set goals and objectives. Thus, the properties manifest to and can be used as procedures, a control mechanism, culture, and communication channels. The properties are influenced by interaction with technology (IT solutions) in an organisation (Orlikowski & Robey, 1991). By implication, IT solutions do more than enable and support business processes in an organisation.

ST helps to detect how a social process comes to involve the interaction of the social structures and agents that comprise various IT specialists in an organisation's environment. This interaction relies on how human actions are enabled or disabled through the power relationship and organisational structures, which are interestingly a result of previous actions (Orlikowski & Robey, 1991).

The actions that are produced and reproduced by actors manifest from DOS (of structure, modality, and interaction) and have intended and

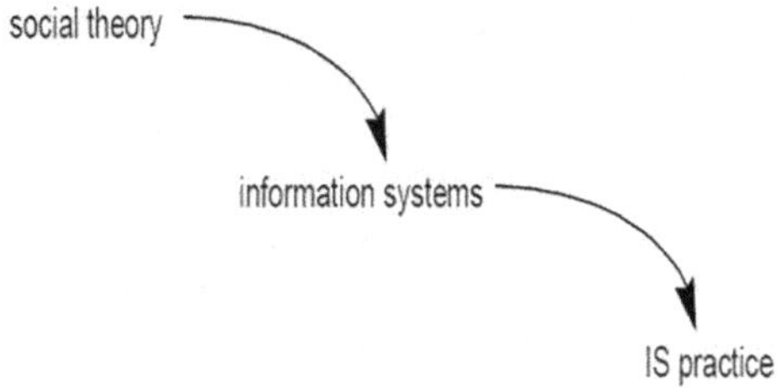

Figure 8.2 ST to ISD practice.
Source: Rose and Scheepers (2001).

unintended consequences. In ST, two types of knowledge – discursive knowledge and practical knowledge – are identified in the course of interaction. The former is the knowledge agents in an organisation are able to articulate and the latter is the tacit knowledge which actors draw upon (Giddens, 1984). Structures are therefore produced and reproduced through the actions of agents (Akgün, Byrne & Keskin, 2007).

ST and IS

ST has been employed in IS studies for over three decades. Rose and Scheepers (2001) attest to the importance ST has in IS (see Figure 8.2). Hence, they have applied the theory and expanded on it to further theorise research in IS. The use of structuration increases the rigour in analysing empirical situations. Also, it clarifies that the relationship between technology and humans must be well understood if any meaningful impact is to be made in an organisation that relies on IT solutions for productivity and sustainability.

ST is a theory founded on social behaviour. In the context of IS, the theory is mostly used to focus on the social aspects of human action to offer a solution to a particular problem. This makes the theory suitable for investigating or examining how employees' actions are influenced by the environment in which they find themselves, as well as the attitudes and responses of others in the same social system.

Although ST was developed within the social context, it is overwhelmingly being theorised in all areas of IS, such as health informatics, software development, and enterprise architecture. A study by Rose (2008) acknowledges the vital role IT plays in organisations. According to Orlikowski (1992), technology is an 'external force' with an impact on the structure of an organisation. Bernardi (2017) applies ST in investigating how health information systems (HIS) influence practices, including the consequences for healthcare provision. The adaptation of agile methodologies in software development was assessed through ST by Cao et al. (2009). Kabanda and Brown (2017) theoretically analysed e-commerce adoption in a developing country by identifying structural practices associated with sociocultural, technological, and contextual settings. According to Orlikowski (1992), ST examines different assumptions, knowledge, and techniques embedded in various kinds of artefacts, including practices and how they can have different consequences for humans and cognitions.

ST is considered to be an independent influence on human behaviour. According to Orlikowski, technology is a result of human action, design, and appropriation. In a model referred to as strategic choice model (refer to Figure 8.3), Orlikowski denounced the assumption that technology is not an external object; it is, instead, a result of human action, design, and appropriation. The model is used to illustrate and argue that technology is dependent on and influenced by the organisation's political processes, the

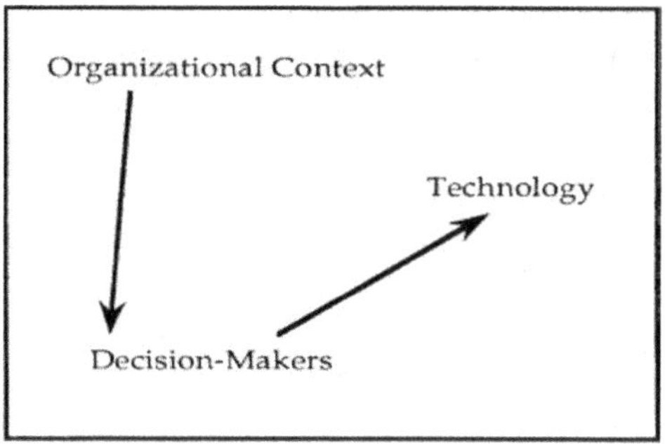

Figure 8.3 Strategic choice model.
Source: Orlikowski (1992).

social environment, and the actions of the employees in the organisations (Pozzebon & Pinsonneault, 2005). Thus, Orlikowski (1992) recognises the vital role people play in the application of IT solutions as they make a decision (see Figure 8.3) in organisations.

The relationship and impact of IS on people and on the organisation is articulated by Sauer and Reich (2009). From the study, it clearly reveal that IT solutions cannot be viewed: (1) from one angle with an organisation, it depends on other structural properties; and (2) in isolation from people or human actions at various levels of an organisation. I emphasise that IS and technologies solutions and humans cannot be viewed in isolation; the two are mutually dependent. From an investigation by Hussain and Cornelius (2009), people are either involved in the deployment of IT solution or indirectly affected.

The *duality of technology* aims to eliminate the dichotomy between the objective view of technology as hardware, equipment, machines, and instruments (Barley, 1986) and the social view of technology. Technology has a physical component as a material artefact and a social component constituted of the different meanings that actors attach to it and the features they emphasise and apply. Its sociological theorising and analysis offer a rich understanding of social action that informs IS and technologies practices towards success or failure. Consequently, some of the decisions or actions are either consciously or unconsciously exhibited.

Applying ST as a lens in IS research

ST is an underpinning theory where an agent is assumed to be the primary entity in a social environment (Giddens, 1984). According to Giddens, the agents in the social system are aware of not only their actions but also the reasons for the decisions they take. Structuration implicitly states that the agents' actions are conditioned by the organisational properties in social contexts (Iyamu & Roode, 2012).

ST is required in information studies to analyse or interpret data in achieving objectives such as the interplay between technical and non-technical factors; the interactions that take place in either the development or implementation or both, and how those interactions manifest themselves; and from enabling and constraining perspectives, how humans and technologies influence and impact social environments. A study underpinned by ST means that the theory was used to guide the analysis or interpretation of the data. In many studies, DOS is applied as a lens.

The trajectory and main tasks of IS and technologies are design, development, and implementation, including post-implementation. In the execution of these tasks, processes and procedures are followed; resources (human and technologies) are allocated; time schedules are drawn up; costs are projected; and budgets are approved. These tasks are executed by individuals and groups in organisations. Agency, according to Rose and Scheepers (2001), has the ability to make a difference. A difference can be achieved by drawing upon power. Power can be used by individuals, depending on how the resources are leveraged in the process of producing IT solutions in an organisation.

From the dimension of analysis, applying DOS as a lens helps in empirical analysis in a detailed form. This can be done at both macro and micro levels of analysis to inextricably link recursive actions and social practice. In the use of DOS from the perspective of structuration, I provide a guide, as shown in Table 8.1. The term 'DOS' in the table refers to duality of structure. The guide presents fundamental questions, using the following interrogatives What, Who, Why, How, When, and Where, to elicit information from participants. The interrogatives help with the analysis of the scheme of things. The answers should be in sequential order and have logical flow. This is intended to unveil and reveal factors on why things happen in certain manner. Also, the answers reveal how events manifest themselves over time and space.

The discussion that follows should be read with Table 8.1 in order to gain a deeper understanding in context and perspective in applying DOS for the analysis or interpretation of data.

Signification/interpretive schemes/communication

Agents in an organisation rely on the available communication channels to execute their responsibilities in the activities of IT solutions. This is critical in that how and what significance an agent associates with instructions or messages is based on how it is communicated, which is influenced by 'how' the agent receives, interprets, and understands it. Also, the 'how' manifests from interest, which is either personal or organisational, or both. Giddens (1984) attributes such behaviour (interest) to the traces of the mind, an act of

Table 8.1 Guide for data analysis

DOS	Signification	Domination	Legitimation
Structure	What makes an IT solution significant? Who considers an IT solution significant? Why is the IT solution considered significant? How is an IT solution considered significant? When (at what stage) is an IT solution considered significant? Where (from what point) is an IT solution considered significant?	What denotes dominance in an IT solution? Who are the actors that dominate in an IT solution? Why do those actors dominate? How is the dominance exerted? When does the dominance reflect or take place? Where exactly does the dominance reflect or take place in an IT solution?	What constitutes legitimate action in an IT solution? Who legitimises activities or actions relating to IT solutions? Why are certain activities or actions relating to IT solutions considered legitimate? How are activities or actions relating to IT solutions legitimised? When do the activities or actions become legitimised? Where are the activities or actions relating to IT solutions legitimised?
	Interpretation	**Facility**	**Norm**
Modality	What aspect of an IT solution requires interpretation? Who carries out the interpretation? Why are interpretations on IT solutions carried out by certain actors? How are interpretations on IT solutions carried out? When are the interpretations on IT solutions carried out? Where within an IT solution are the interpretations carried out?	What facilities exist for IT solutions deployment? Who are the actors responsible for applying the facilities in the deployment of IT solutions? Why are those actors responsible and why were the facilities employed in the manner that they were? How are facilities employed in IT activities or processes of IT solutions? When are facilities applied in the activities or processes of IT solutions? Where are facilities applied in the activities or processes of IT solutions?	What are factors considered norms in the deployment of IT solutions? Who has the power to endorse an activity as a norm? Why are certain activities of IT solutions endorsed as norms? How are certain activities of IT solutions endorsed as norms? When in the deployment of IT solutions are certain activities endorsed as norms? Where in the deployment of IT solutions are certain IT solutions endorsed as norms?

(*Contined*)

	Communication	*Power*	*Sanction*
Interaction	What aspects of IT solutions are communicated? Who carries out the communication, and to whom? Why are certain communications about IT solutions carried out in the manner that they are? How is the communication about IT solutions carried out? When is the communication about IT solutions carried out? Where is the communication carried out?	What constitutes power in the activities of IT solutions? Who amongst the actors have the power to make a decision in the activities of IT solutions? Why do some actors have power over others in the activities of IT solutions? How is power acquired and used in the activities of IT solutions? When is power used in the activities of IT solutions? Where does the power come from and where is it exerted within a system?	What activities are sanctioned in the deployment of IT solutions? Who sanctions activities or processes in the deployment of IT solutions? Why are certain activities or processes sanctioned in the deployment of IT solutions? How are activities or processes sanctioned in the deployment of IT solutions? When are activities or processes sanctioned in the deployment of IT solutions sanctioned? Where are activities or processes sanctioned in the deployment of IT solutions sanctioned?

unconsciousness that comes from sentiments of the self-hood. In his explanation of how one relates to self, an individual's consciousness is the layer upon which an individual may either know or claim to process knowledge (Giddens, 1987).

Communication is vital to actors, as they tap into their interpretive schemes to conceptualise the design, development, implementation, and use of IT solutions for various reasons within an environment. The importance of communication in an activity such as deployment of IT solutions lies in the way in which an individual understands it and can have success or failure bearing on the outcome.

Organisational structures can be affected by individuals' interpretive schemes and structures. Agents interpret and internalise information communicated to them and, from this, they are able to act accordingly. Sometimes, agents take actions that they may not be able to justify or explain mainly because it has become a norm with them. Also, there are certain things that an individual might not be able to express in words but could communicate in other ways. This type of action is determined or influenced by the available resources, rules, or regulations within an organisation.

The primary purpose of a communication strategy in an IT project is to communicate its activities and processes with the stakeholders. The communication can be carried out in different ways, such as in written documents and meetings. When working on a project, some agents carry out certain activities or tasks that they cannot explain. They do these things because they just know how to do them. This can often be attributed to the fact that they had been doing it for so long that it had is self-inscribed and become a part of them.

Domination/facility/power

Agents (employees) are able to use their influence to get favourable results, swing decisions of IT solutions' deployment in their way for personal or organisational interest, or both. The agents do so, wielding their power or using resources they obtain from the organisation. Sauer and Reich (2009) highlight and emphasise the importance of control of activities in order to achieve success. Hussain and Cornelius (2009) suggest that power is a phenomenon used to control resources in an environment. Thus, the facilities used in obtaining power are not restricted but can be limited by other actors. One of the facilities is knowledgeability. Such individuals are referred to as knowledgeable agents. It is the reflexive form of the knowledgeability of human agents that is involved in recursive practices in a social system.

Agents exercise power in various ways in the course of developing, implementing, or managing IT solutions. This implies that individuals involved in activities of IT solutions have exponential degree of power. Assigning an individual to deployment of IT solutions implies that they have the ability to make a difference and influence the outcome of the IT solution. This means that every agent has a degree of knowledge or upholds resources that qualify them as stakeholders, and which give them the power to make a difference. Agents are identified as stakeholders because they control resources (Hussain & Cornelius, 2009).

Legitimation/norm/sanction

An agent creates the structure of legitimation through norms. Agents sanction (endorse) each other's actions by drawing on norms, regulations, or standards of value and morality as deemed acceptable in an organisation. Organisational structure determines how employees process information which they consider acceptable in carrying out activities of IT solutions. Consequently, the information determines how the IT solutions are developed, implemented, managed, or used within the norm and value of an organisation to promote sustainability and competitiveness. Agents draw on the vision and requirements of the organisation to legitimise their own actions or decisions to select or implement an IT solution.

There seem to be different schools of thought in making sense of Giddens and structuration. Rose (1998) explains how structuralists and functionalists put meaning to structure through social behaviour and structural forces, which encourage people to behave in a certain manner. Although there was no focus on structuration in isolation, they have explored and demonstrated how communication can change once an organisation has adopted new ways. The culture changes to suit the new way of doing things. From a similar perspective, Orlikowski (1992) clarifies how structuration can be applied to examine how the rules and procedures in a system can be changed and carried out by the agents in their respective work.

From acceptance of agents' actions, culture is created and becomes the norm in an organisation. Based on the culture, activities and processes are carried out to select, develop, implement, or use IT solutions for an organisation's purposes. However, the same norm is not applied to all activities, such as selection, development, implementation, or practice of IT solutions. Hence, the questions used to elicit information from the participating actors are slightly different. Also, different groups of employees or units in an organisation adopt different ways of doing things within rules and regulations that are sanctioned by the organisation. Organisational culture is the way an organisation deploys its IT solutions towards achieving its goals and objectives.

Conclusion

This chapter unpacks and discusses the tenets of ST in a simple manner. The chapter presents how ST is used to explain interactions between agents and structure in the activities and processes of IT solutions. This is done through

DOS, using structure, modality, and interaction to elicit information concerning activities of IT solutions in organisations. From the guide (Table 8.1), it shows how power, particularly control of resources, is acquired and enacted to carry out activities and processes that influence the selection, development, implementation, or practice of IT solutions in organisations.

The chapter presents a practical guide which can help researchers and students in their quest to employ ST as a lens in IS studies. As demonstrated in the chapter, ST is useful in IS research to provoke the interplay between the accounts of agents in the deployment of IT solutions. Through the accounts, assessment and evaluation can be conducted to gain insights on the effects of the actions of individuals interacting with structures (rules and resources) in the activities or processes of IT solutions.

In applying the guide, one needs to take cognisance of three fundamental facts: (1) that organisational transformation (from the development to implementation of IT solutions) is the joint effect of the actions of individuals and groups interacting within institutional structures; (2) structures enable and constrain the daily actions and thought processes of people, but do not wholly determine the results; (3) individuals' choices are not independent of the structures within which agents operate but they can move towards impacting (maintaining, reinforcing, changing, or even destroying) the results.

References

Akgün, A. E., Byrne, J., & Keskin, H. (2007). Organizational intelligence: A structuration view. *Journal of Organizational Change Management, 20*(3), 272–289.

Algesheimer, R., & Gurău, C. (2008). Introducing structuration theory in communal consumption behavior research. *Qualitative Market Research: An International Journal, 11*(2), 227–245.

Barley, S. R. (1986). Technology as an occasion for structuring: Evidence from observations of CT scanners and the social order of radiology departments. *Administrative Science Quarterly, 31*(1), 78–108.

Bernardi, R. (2017). Health information systems and accountability in Kenya: A structuration theory perspective. *Journal of the Association for Information Systems, 18*(12), 931–958.

Cao, L., Mohan, K., Xu, P., & Ramesh, B. (2009). A framework for adapting agile development methodologies. *European Journal of Information Systems, 18*(4), 332–343.

Falkheimer, J. (2007). Anthony Giddens and public relations: A third way perspective. *Public Relations Review, 33*(3), 287–293.

Furumo, K., & Melcher, A. (2006). The importance of social structure in implementing ERP systems: A case study using adaptive structuration theory. *Journal of Information Technology Case and Application Research, 8*(2), 39–58.

Gao, P., & Lyytinen, K. (2003). China telecommunications transformation in globalization context. In: M. Korpela, R. Montealegre, & A. Poulymenakou A. (eds), *Organizational information systems in the context of globalization*, pp. 217–233. Boston, MA: Springer.

Giddens, A. (1984). *The constitution of society: Outline of the theory of structuration*. Cambridge: Polity.

Giddens, A. (1987). *Social theory and modern sociology*. California: Stanford University Press.

Hussain, Z. I., & Cornelius, N. (2009). The use of domination and legitimation in information systems implementation. *Information Systems Journal*, *19*(2), 197–224.

Iyamu, T., & Roode, D. (2012). The use of structuration theory and actor network theory for analysis: Case study of a financial institution in South Africa. In: A. Tatnall (ed.), *Social influences on information and communication technology innovations*, pp. 1–19. Hershey: IGI Global.

Iyamu, T., & Shaanika, I. (2020). Factors influencing the use of mobile systems to access healthcare big data in a Namibian public hospital. *Information Resources Management Journal (IRMJ)*, *33*(3), 81–99.

Kabanda, S., & Brown, I. (2017). A structuration analysis of Small and Medium Enterprise (SME) adoption of E-Commerce: The case of Tanzania. *Telematics and Informatics*, *34*(4), 118–132.

Orlikowski, W., & Robey, D. (1991). Information technology and the structuring of organizations. *Information Systems Research*, *2*(2), 143–169.

Orlikowski, W. J. (1992). The duality of technology: Rethinking the concept of technology in organizations. *Organization Science*, *3*(3), 398–427.

Pozzebon, M., & Pinsonneault, A. (2005). Challenges in conducting empirical work using structuration theory: Learning from IT research. *Organization Studies*, *26*(9), 1353–1376.

Rose, J. (1998, June). Evaluating the contribution of structuration theory to the information systems discipline. In *6th European Conference on Information Systems* (Vol. 2, pp. 910–924). Aix-en-Provence, France.

Rose, J., & Scheepers, R. (2001, June). Structuration theory and information system development-frameworks for practice. In *Proceedings of the 9th European Conference on Information Systems* (pp. 27–29). Bled, Slovenia.

Sauer, C., & Reich, B. H. (2009). Rethinking IT project management: Evidence of a new mindset and its implications. *International Journal of Project Management*, *27*(2), 182–193.

9 Technology acceptance model

Table of contents

Introduction

As clarified in the introductory chapter, information systems and technology (IS/IT) and information technology (IT) solutions mean the same thing in the context of this book. Thus, both IT and IS/IT are interchangeably used in this and other chapters. Another important clarification that I would like to make is that environment and organisation mean the same thing. They are also interchangeably used in this chapter.

Human interaction is inseparable from the adoption and use of IT solutions. Consequently, it is always essential to investigate how they became inseparable, the factors of influence, and how the inseparability manifests. Knowledge or assumed know-how are some of the main factors that influence human acceptance or rejection of technology solutions in organisations. In examining the determinant factors of why and how IT solutions are accepted or rejected, the technology acceptance model (TAM) is increasingly applied in IS research. Also, this can be attributed to the fact that the relationships often manifest from assumptions and perceptions about the use of technology, which is not always easy to assess or evaluate. TAM is one of the theories that have been explored and used in information systems (IS) studies, particularly in research that is of sociotechnical nature. Some of the theories are covered from Chapters 4 to 8. TAM is adaptable and concerned with human relationship with technology. TAM provides IS researchers with options of applying it either as a model or as a theory for studying IS/IT solutions' implementation, usage, and management. Human

DOI: 10.4324/9781003184119-9

relationships and interactions with technology have been found to be critical in the development and implementation as well as in the use and management of IT solutions. Thus, applying TAM in IS research has increased over the years.

TAM was introduced over three decades ago by Davis (1985). It is one of the theories more widely applied in IS research to examine and explain IS/IT rejection or acceptance and usage. Davis (1989) explains that the primary aim of TAM is to provide a basis for discovering the impact of external variables on internal beliefs, attitudes, and intentions. TAM assumes that beliefs about usefulness and ease of use are always the primary determinants of IT solutions' adoption in organisations. In TAM, these two determinants serve as the basis for attitudes towards using a particular system (IT solution), which in turn determines the intention to use, and generates the actual usage behaviour. The model's suitability for predicting general individual acceptance is continually explored, particularly with emerging IT solutions. As shown in Figure 9.1, TAM consists of five main elements, namely, perceived usefulness (PU), perceived ease of use (PEoU), attitude towards using, behavioural intention (BI), and actual system use.

BI is in some ways determined by a subjective norm from interpretation of communication which an individual receives. The communication could come from various sources, such as a colleague, employer, or superior, which ultimately manifests to influence a user's behaviour (Ajibade, 2018). Evidently, individuals' perceptions of IT solutions are likely to be influenced by the objective characteristics of technology, as well as interaction with other users (Davis, 1989; Venkatesh & Morris, 2000).

The introduction of new technology does not necessarily mean automatic acceptance by potential users. TAM is primarily used to explain why and how users come to accept and make use of a specific technology over time and space (Louho, Kallioja & Oittinen, 2006). Technology acceptance is mainly about why and how people accept and use the introduced technology (Bertrand & Stephane, 2008). The model is suitable in the analysis and interpretation of emerging technical issues in the implementation of new technologies in particular, within an environment or case. Primarily, TAM has been developed for system use, acceptance, and user satisfaction of

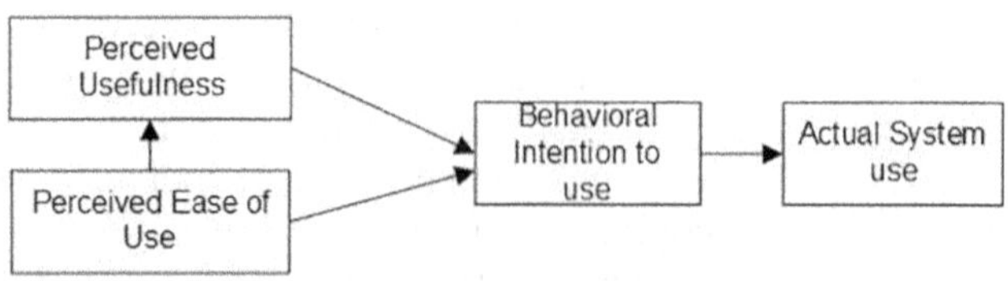

Figure 9.1 Technology acceptance model.
Source: Davies et al. (1989).

systems (IT solutions) (Davis, Bagozzi & Warshaw, 1989). The theory is an attempt to simplify and make easier the prediction of users' acceptance of technology solutions, which is critically important in organisations.

However, applying TAM has not always been easy or straightforward, particularly for postgraduate students. This is a consequence of not having a formula or acceptable pattern for applying the theory. The challenge is synonymous in applying sociotechnical theories including those covered in previous chapters. This chapter proposes an easier method of applying TAM in IS research. In doing so, first, the chapter is sequentially structured into five main sections. First, is the introductory section, which provides explanation about the significance, as well as the challenges of applying the theory. This is followed by a review of literature on TAM. In the third section, TAM is covered in the context of the IS field. A method of applying TAM in IS research is proposed in the fourth section, and the chapter is concluded in the fifth.

Technology acceptance model

TAM is a behavioural model based on Ajzen and Fishbein's (1980) theory of reasoned action (TRA). TAM is one of the effective contributions of TRA. Davis (1989) developed TAM to explain and predict users' acceptance of IS/IT solutions. It aims to determine the factors that influence human decisions about the technology that they reject or accept and use in organisations. In other words, TAM intends

> to provide an explanation of the determinants of computer acceptance that is general, capable of explaining user behaviour across a broad range of end-user computing technologies and user populations, while at the same time being both parsimonious and theoretically justified.
>
> (Davis et al., 1989)

TAM models how users come to reject or accept and use an IT solution. It is used to facilitate an understanding of relationships that exist between external variables that affect the acceptance of IT solutions by users. Consequently, the theory hypothesises a relationship between external variables and both PU and PEoU. Thus, TAM is often considered a suitable model (or theory) to underpin IS research that is of predictive nature because it explains and gives the understanding of factors that have an impact on acceptance or rejection of IT solutions in organisations. It is on this basis that TAM has been employed from a social context perspective in order to examine and comprehend individuals' and groups' perceptions in the adoption and use of IT solutions for organisational services.

Another important aspect raised by Bagozzi (2007) is the gaps which exist between intention-behaviour linkages that suggest that many considerations could take place between users' intention to perform a certain action and performing the actual action. In this chapter, three considerations are highlighted as follows: (1) actions are taken as a means to an end (the adoption of technology is to facilitate a function in an organisation); (2) intention precedes action, with many intervening steps in between that could impede eventual action; and (3) decisions are made on the basis of obstacles and impediments and by trying to accept a technology a change of focus can occur and probably influence the decision-making process. As noted by Venkatesh and Davis (2000), the actions of an individual can be influenced by people who are important to that individual. With this in mind, Bagozzi (2007) highlights the influential power of group, cultural, and social pressures, stating that the decision to reject or adopt a technology is in many ways collaborative in nature.

TAM is increasingly being applied in IS research to explore the acceptance of new IT solutions or new e-services (Davis, 1989; Davis & Venkatesh, 1996). As shown in Figure 9.1, TAM is built upon PU and PEoU. With regard to PU, Davis (1989) states that people tend to use or not use an application to the extent that they believe it will assist them to achieve their objectives or perform certain responsibilities. The conscious and unconscious thoughts that precede are that even when a potential user envisages that an IT solution is useful, he or she may at the same time believe that the system is too hard to use and that the performance benefits of usage are outweighed by the effort of using it. Davis et al. (1989) define the two variables as follows:

i PU is 'the degree to which a person believes that using a particular system would enhance his or her job performance'.
ii PEoU is 'the degree to which a person believes that using a particular system would be free of effort'.

Figure 9.1 illustrates the four main components of TAM. It is a foundation for a discourse, analysis, or interpretation of the trajectory of why and how humans come to reject or accept IT solutions in organisations.

TAM establishes a theoretical basis for explaining causal links between the two key constructs, PU and PEoU, of an ISsolution, and users' attitudes towards use (AU); BI; and the actual use of the solution (AS). Acceptance is the highest-level concept and is a function of perceived utility, usability, likeability, and cost. Utility refers to the match between a user's needs and the product's functionality. Usability refers to users' utilisation of a product's functionality in practice. Likeability refers to affective judgements and costs which are the financial, social, and organisational effects that are associated with IT solutions. Usability dimensions include:

i Effectiveness: the results from interaction within a certain speed and accuracy or errors.
ii Learnability: the relationship of training and frequency of use and performance.

iii Flexibility: the adaptation to some specified tasks and/or environments beyond first specified.
iv Attitude: the user-acceptable level of human cost in terms of tiredness, discomfort, frustration, and personal effort. This dimension of usability is explicitly dependent on users' opinions.

System acceptability is divided into two main social and practical categories. Practical acceptability is influenced by usefulness, costs, compatibility, and reliability. Usefulness is further split into two: utility and usability. The first aspect, utility, is whether the functionality of the system in principle can perform the tasks that are needed. The other aspect, usability, focuses on how well users can make use of the functionality. User acceptance is defined by Dillon and Morris (1996) as the demonstrable willingness within a user group to employ an IT solution for the tasks it is designed to enable and support. By indication, this definition focuses on planned and intended uses of technology. However, studies report that an individual's perceptions of IT solutions are likely to be influenced by the objective characteristics of technology, as well as interaction with other users (Davis, 1989; Venkatesh & Morris, 2000).

IS in an institutional setting make IT policies and rules take precedence over user attitudes. This indicates that the guiding rules of the organisation serve as influence and controlling factors on user behaviour (Ajibade, 2018). The users' perceptions of an IT system influence acceptance, and how individuals adopt these new systems has long been an active area of research (Venkatesh et al., 2003). Technology acceptance theories explain that user intentions and acceptance of IT solutions are broadly based on IS, behavioural sciences, and social psychology theories that describe how individuals perceive technology and make a decision to adopt or not adopt the new technology (Moon & Kim, 2001; Venkatesh et al., 2003).

TAM in IS research

IT solutions' development and implementation, including use and practice, can be costly and they have had a relatively low success rate in many organisations. However, there have been some improvements over the years. Legris, Ingham, and Collerette (2003) argue that IS research has contributed to a better understanding of these processes and their outcomes. The development and implementation of systems and technologies will continue to improve as different theoretical approaches, such as TAM, are applied to promote awareness and gain a better insight of why and how various factors influence acceptance or rejection of technology from social context perspectives. Marangunić and Granić (2015) suggest that external variables and potential system usage (PSU) are also considered when evaluating technology acceptance or usage. TAM is thus most suitable for predicting technology rejection or acceptance and use of a new or existing technology in an environment.

Generally, TAM attempts to address why users accept or reject technology solutions and how users' acceptance is influenced by system characteristics. According to Legris et al. (2003), over the years, TAM has proven to be a theoretical model which can be applied to explain and predict users' behaviour towards IT solutions in various environments. This stems from the fact that TAM assumes that people tend to use or not use technology solutions to the extent that they believe it will help them perform their jobs in better ways. In Park's (2009) view, TAM provides a basis that can be applied to trace why or how external variables influence belief, attitude, and intention to use IT solutions. As depicted in Figure 9.1, TAM consists of four main components, namely, PEoU, PU, BI, and actual system use.

Based on the preceding review, it is evident that the theory focuses on three main areas, which are: (1) to examine and understand the adoption and use of IT solutions; (2) to gain insights on why or how individuals' and groups' perceptions affect the intentions to use IT solutions from different perspectives; and (3) to create comprehension of why or how IT solutions can be accepted and used for its intended purposes. According to Udo and Bagchi (2011: 4), 'the independent constructs of TAM are perceived ease of use and perceived usefulness, while the dependent constructs are behavioural intention to use and system usage'.

My understanding of TAM, relative to IS research, leads to the development of Figure 9.2. As shown in the figure, the intention is to demonstrate how the elements are linked together to influence or depend on each other. First, it shows the essentiality of external variables. Second, it explicates how the BI to use and system usage are influenced by the organisational needs, as well as the requirements that guide the system itself. The figure therefore helps to create the fundamental basis on which the trajectories (why and how) of rejecting or accepting IT solutions in an environment can be analysed or interpreted.

To an organisation, in terms of needs, usefulness means achieving goals and objectives through the enablement and support of IT solutions. Achieving goals and objectives is often measured by increases in productivity and return on investment as dictated by requirements. The organisational needs are derived from requirements from both the technical and non-technical perspectives. Also, the requirements guide and determine individuals' use of IT solutions to enable and support organisational needs. According to Davis (1989), TAM presumes that PU and PEoU determine an individual's intention to actually make use of a system, with the intention to use serving as a mediator of actual system use.

As shown in Figure 9.2, TAM is applied in IS research based on objectives that are purposely to investigate how to improve the services of an organisation through an understanding of IT solutions' acceptance. The acceptance of a new IT solution has been shown to be a prerequisite for successful implementation, and also its determinants of user's acceptance can contribute to enhancing system design and use (Agarwal & Prasad, 1998; Davis,

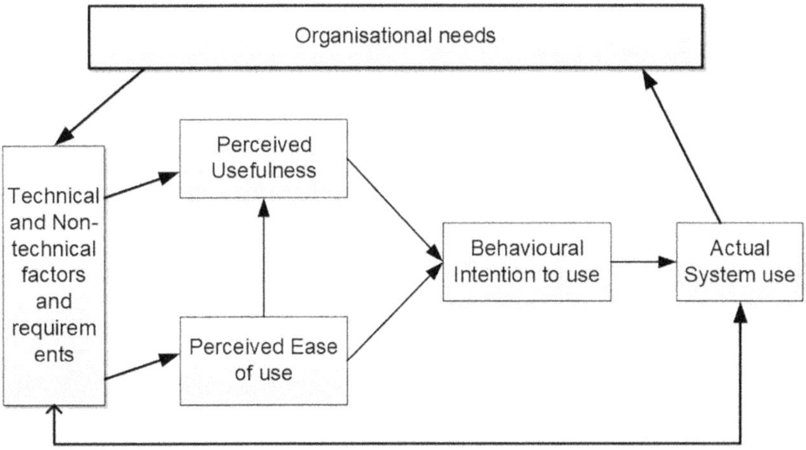

Figure 9.2 TAM in IS.
Source: Author.

1989). Another critical aspect is examining how users' perceptions influence acceptance of IT solutions and how individuals adopt new systems, which have long been active areas of research in IS (Venkatesh et al., 2003). Subsequently, TAM is considered a useful theory to underpin IS studies in that it explains and assists to gain in-depth knowledge of the factors that impact on the acceptance of IT solutions in organisations.

TAM for analysis in qualitative studies

TAM is used for evaluating and determining the users' behaviour and acceptance of IT solutions by determining their PU and PEoU (Lee, Kozar & Larsen, 2003). It is thus used for analysis and interpretation, specifically, of qualitative data, as shown in Figure 9.3. TAM is considered a theory of reasoning in that it has strong behavioural elements and assumes that when someone develops an intention to act, they can freely act without limitation. In essence, it is necessary to point out that in practice, constraints such as limited ability, time, environmental or organisational limits, and unconscious habits do limit the freedom to act. Hence, it is essentially important to employ the interrogatives 'who, how and why' in probing responses during data collection in IS research, using TAM.

TAM helps to uncover how external variables contribute to influencing users' acceptance and actual use of IT solutions (Yang & Wang, 2019). External variables include social influence in the use of adopted technologies by users (Taherdoost, 2018). It is this sociotechnical coverage that motivates the wide application of TAM in IS research for analysis and interpretation,

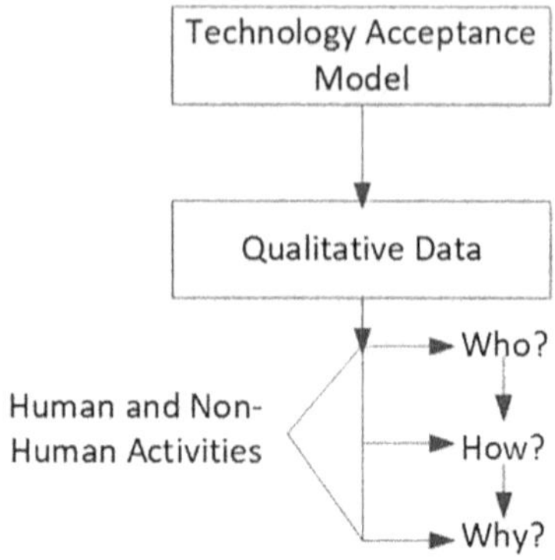

Figure 9.3 TAM for analysis of qualitative data in IS research.
Source: Author.

in predicting users' rejection or acceptance and use of IT solutions in organisations (Ahmad, 2018; Al-azawei, Parslow & Lundqvist, 2017; Guner & Acarturk, 2020). Significantly, TAM contributes to gaining insights into who, how, and why individuals reject or make use of IT solutions for various purposes (Surendran, 2012).

Even before engagement, a system can be perceived to be useful within a context, based on the subjective view of the intended user. This raises three fundamental questions: (1) Who perceived the system useful, other than the focal actor? (2) Why was the system perceived useful in the context that it was? (3) How was the system perceived useful? These questions help to explore the rationale and understanding of the PU of a system. PU is conceived and based on perception, informed by subjectivity. Subjectivity is an expression of know-how which has no particular form of measurement to be correct or incorrect. Hence, individuals' or groups' perception of usefulness of a system is neither right nor wrong. But it does lay a foundation for the analysis, interpretation, and explanation of why and how an IT solution is adopted and used in an environment.

The PEoU of a system is situational and experimental, resulting in subjective experience and opinion. This occurs over a period of time. Such a circumstance raises important questions: Why would a system be perceived as of 'ease of use' if it has not been used? Who perceived the system to be of 'ease of use' contextually? The decision or proposition to perceive a system to be of 'ease of use' is informed or based upon opinions of an individual or group of actors at the time, which is subject to change as the circumstances unfold. Thus, Table 9.1 is designed to help facilitate applying TAM in IS research for analysis or interpretation of qualitative data.

Table 9.1 TAM for data analysis

Component	*Who*	*How*	*Why*
PU	Identify the functionalities of the IT solution. Identify the users at different levels and units that think the IT solution worthwhile.	Establish the criteria (such as business and technical objectives and requirements) for thinking an IT solution will be of value and suitable in an organisation.	Determine the rationale in the form of merit or demerit for considering an IT solution as a useful or not useful utility in an environment.
PEoU	Identify the factors that make the users envisage that an IT solution can be easy to use in their organisation.	Examine the benchmarks that inform the thoughts that were employed by users that an IT solution is of ease of use.	Corroborate the factors that influence the decision or thoughts that an IT solution can be or is of ease of use.
BI to use	At different levels and units, identify and validate the factors that influence users' behaviour in their various intentions to use an IT solution.	Substantiate the rationale or factors that influence actors wanting or being interested in using a particular IT solution. Understand the consequence of the plan to make use of an IT solution.	Establish the varying interests of the actors, whether it is voluntary or obligatory, and determine the motivating factors of influence for wanting to use a particular IT solution.
Actual system use	Verify the users of the IT solutions and identify the aspects of the system that are being put to use.	Determine the different methods or approaches that are employed in using the IT solution or parts of the system.	Determine the rationale for using the IT solution at the different levels or units by various actors.

As already established or argued, PU and PEoU are based on perception, which does not completely eradicate bias. Thus, actors are influenced by their stock of knowledge of the IT solutions gathered over time. The stock of knowledge is gathered from experience, training, or education through appetite for learning and self-inquisitiveness.

Behaviour is subjective; as such, no theory can positively account for human actions as a whole. Human behaviour is an enactment of power acquired from different sources and used with the intention to achieve an objective. The challenge remains in the attempt to examine the intentional act without understanding the source of power. However, power is not easily explanatory because power in itself can be of practical unconsciousness. Giddens (1984) described practical unconsciousness as entailing what people do in everyday life but may not be able to put into words. Hence, it is empirically essential to investigate, using the interrogatives who, how, and why to elicit information from participants in IS research.

Conclusion

In TAM, PEoU is regarded as a process of expectancy, primarily because it focuses on a person's subjective perception about the level of effort which is needed to use an IT solution. It therefore represents users' subjective evaluation and assessment of activities and processes that are involved in interactions with IT solutions (Davis & Wong, 2007). PU is the extent to which a person thinks or believes the use can be of significant value or suitable in performing or improving his or her processes and activities, including employment responsibilities. Usefulness in this context is the extent to which individuals and groups of users perceive the systems to be useful in getting their work done.

Actual system use refers to users' utilisation of the system's functionalities. Any IS that is not used has little or no value in it. This is also a perception that needs empirical evidence examined through TAM. Considering investment and competitiveness, it is essential to examine, when practical to do so, why employees choose to use or not to use an IT solution in their organisations. In many IS researches, TAM is employed primarily to examine factors which affect and influence users' acceptance of technologies and systems in organisations. It helps to determine how users form attitudes towards rejection or acceptance of IT solutions and services in organisations, based on the characteristics.

References

Agarwal, R., & Prasad, J. (1998). The antecedents and consequences of user perceptions in information technology adoption. *Decision Support Systems*, *22*, 15–29.

Ahmad, M. (2018). Review of the Technology Acceptance Model (TAM) in internet banking and mobile banking. *International Journal of Information Communication Technology and Digital Convergence*, *3*(1), 23–41.

Ajibade, P. (2018). Technology acceptance model limitations and criticisms: Exploring the practical applications and use in technology-related studies, mixed-method, and qualitative researches. *Library Philosophy & Practice*, *1941*, 1–13.

Ajzen, I., & M. Fishbein (1980). *Understanding attitudes and predicting social behaviour.* Upper Saddle River, NJ: Prentice-Hall.

Al-Azawei, A., Parslow, P., & Lundqvist, K. (2017). Investigating the effect of learning styles in a blended e-learning system: An extension of the Technology Acceptance Model (TAM). *Australasian Journal of Educational Technology*, *33*(2), 1–23.

Bagozzi, R. P. (2007). The legacy of the technology acceptance model and a proposal for a paradigm shift. *Journal of the Association for Information Systems*, *8*(4), 244–254.

Bertrand, M., & Stephane, B. (2008). Applying the technology acceptance model to VR with people who are favourable to its use. *Journal of Cyber Therapy & Rehabilitation*, *1*(2), 200–210.

Davis, F. D. (1985). *A technology acceptance model for empirically testing new end-user information systems: Theory and results.* Doctoral dissertation, Massachusetts Institute of Technology.

Davis, F. D. (1989). Perceived usefulness, perceived ease of use, and user acceptance of information technology. *MIS Quarterly*, *13*(3), 319–340.

Davis, F. D., Bagozzi, R. P., & Warshaw, P. R. (1989). User acceptance of computer technology: A comparison of two theoretical models. *Management Science*, *35*(8), 982–1002.

Davis, F. D., & Venkatesh, V. (1996). A critical assessment of potential measurement biases in the technology acceptance model: Three experiments. *International Journal of Human-Computer Studies*, *45*(1), 19–45.

Davis, R., & Wong, D. (2007). Conceptualizing and measuring of the e-learning environment. *Decision Science Journal of Innovative Education*, *5*(1), 982–1002.

Dillon, A., & Morris, M. (1996). User acceptance of new information technology: Theories and models. *Annual Review of Information Science and Technology*, *31*, 3–32.

Giddens, A. (1984). *The constitution of society: Outline of the theory of structuration.* Cambridge: Polity.

Guner, H., & Acarturk, C. (2020). The use and acceptance of ICT by senior citizens: A comparison of technology acceptance model (TAM) for elderly and young adults. *Universal Access in the Information Society*, *19*(2), 311–330.

Lee, Y., Kozar, K. A., & Larsen, K. R. (2003). The technology acceptance model: Past, present, and future. *Communications of the Association for Information Systems*, *12*(1), 752–780.

Legris, P., Ingham, J., & Collerette, P. (2003). Why do people use information technology? A critical review of the technology acceptance model. *Information & Management*, *40*(3), 191–204.

Louho, R., Kallioja, M., & Oittinen, P. (2006). Factors affecting the use of hybrid media applications. *Graphic Arts in Finland*, *35*(3), 11–21.

Marangunić, N., & Granić, A. (2015). Technology acceptance model: A literature review from 1986 to 2013. *Universal Access in the Information Society*, *14*(1), 81–95.

Moon, J., & Kim, Y. (2001). Extending the TAM for a world-wide-web context. *Information & Management*, *38*(4), 217–230.

Park, S. Y. (2009). An analysis of the technology acceptance model in understanding university students' behavioural intention to use e-Learning. *Educational Technology & Society*, *12*(3), 150–162.

Surendran, P. (2012). Technology acceptance model: A survey of literature. *International Journal of Business and Social Research*, *2*(4), 175–178.

Taherdoost, H. (2018). A review of technology acceptance and adoption models and theories. *Procedia Manufacturing*, *22*, 960–967.

Udo, G. J., & Bagchi, K. K. (2011). Understanding the influence of espoused culture on acceptance of online services in a developing country. *Journal of Information Technology Theory Application*, *12*(2), 25–46.

Venkatesh, V., & Davis, F. D. (2000). A theoretical extension of the technology acceptance model: Four longitudinal field studies. *Management Science*, *46*(2), 186–204.

Venkatesh, V., & Morris, M. G. (2000). Why don't men ever stop to ask for directions? Gender, social influence, and their role in technology acceptance and usage behaviour. *MIS Quarterly*, *24*(1), 115–139.

Venkatesh, V., Morris, M. G., Davis, G. B., & Davis, F. D. (2003). User acceptance of information technology: toward a unified view. *MIS Quarterly*, *27*(3), 425–478.

Yang, Y., & Wang, X. (2019). Modeling the intention to use machine translation for student translators: An extension of Technology Acceptance Model. *Computers & Education*, *133*, 116–126.

10 Selecting theories to underpin information systems research

Table of contents

Introduction

Information systems and information technologies (IS/IT) are aspects of computing that continue to involve more of social contexts in their development, implementation, and practice. The social context makes it more interestingly challenging when it comes to the areas of research. Researches in IS are carried out primarily to investigate and understand what factors are needed; why certain processes are applied; how activities are executed towards achieving aims and objectives; and the impact and influence which the processes, events, and activities have on the development, implementation, and use of information technology (IT) solutions in environments. Thus, sociotechnical theories are increasingly applied in information systems (IS) research to expand the depth of assessments, evaluations, and investigations. Some of the theories are covered in this book from Chapters 4 to 9.

Sociotechnical theories are forms of new streams that offer fresh insights and perspectives into the ways IS/IT mediate in almost all situations of modern organisations. On the academic front, theories add rigour to research. According to Doolin (1998), the factors that influence the practices of IT solutions need to be analysed in a wider context involving social and power relations. Thus, sociotechnical theories are increasingly applied to underpin research in the IS field. Despite this advancement and impressive contributions to the IS body of knowledge, there are two critical questions of fundamental importance to researchers and postgraduate students: (1) what makes a theory suitable or most appropriate for an IS research? (2) what is

DOI: 10.4324/9781003184119-10

the significance of complementarity of theories in a research? These are the questions that this chapter addresses. The latter is an evident indication that there is no one theory that fits best in many of the studies.

In the last two decades, there has been an overwhelming increase in applying sociotechnical theories to underpin IS studies. However, the selection of the theories has been a cumbersome process for many researchers, particularly early career researchers, including postgraduate students. As a result of the complexity, some postgraduate students select the theories that they are comfortable with, as opposed to suitability for their studies. This makes some of their studies more complex, and sometimes delays completion. Also, the challenges with selection influence some postgraduate students to avoid applying sociotechnical theories even though they would have added rigour and quality to the studies. The problem with selection of theories is critical in that, when applied, they shape, influence, and define the results of the studies.

In order to select a theory or theories to underpin a study, the researcher or student should have a good understanding of the theories and the factors that influence the selection. However, many researchers, particularly aspiring researchers in the field of IS, are challenged in applying one theory or the other. In the process of the challenges, two things emerge: Some researchers or postgraduates do not understand the theories well enough to select a suitable one. Some researchers or postgraduates that have a good understanding of some of the theories are often challenged in applying them. The challenges become more cumbersome when two or more theories have to be complementarily applied in a study.

Even though sociotechnical theories have been applied hundreds of thousands of times in IS research, it is hard to find literature or books that offer a guide, model, or any form of standardised approach for selecting sociotechnical theories for IS research. This is critical because the theory shapes the outcome of the research where it is applied. Therefore, there must be a clear rationale for the choice of a theory to ensure that it aligns with and supports the structure of the study. According to Lim et al. (2013), theories are used to provide guidance on analysis and explanation, and to predict the outcome of phenomena. The theories are used to assist students and researchers of IS to develop knowledge that contributes to the body of knowledge.

Some of the sociotechnical theories that have been more commonly applied in IS studies are activity theory (AT), actor-network theory (ANT), contingency theory (CT), diffusion of innovation (DOI) theory, structuration theory (ST), and the technology acceptance model (TAM). Despite their increasing popularity, selecting the theories to underpin IS studies has never been easy (Sawyer & Jarrahi, 2014). As detailed in Chapters 4–9, the theories bear different focuses and offer corollaries in applying them as lenses to analyse and interpret data in IS research.

The selection of theories to underpin IS studies is primarily influenced by the objectives of the study. The way in which phenomena are studied and interpreted is influenced and shaped by the theories that are often employed. These theories are used as guidelines for data collection, analysis of data, and interpretation of the findings from the analysis. The objectives are normally aligned with the research problem, research questions, as well as the research topic. The objectives of the research are to solve a research problem; the research questions are used in eliciting responses and contributions to solve the problem; and the topic guides the focus of the study. This makes the selected theory to be fully embedded in the study in entirety.

This chapter is divided into five main sections. First is the introductory section, followed by a section that covers IS. In the third section, the role of methodology in applying sociotechnical theory is discussed. A proposed approach on how to select theories is presented in the fourth section. Thereafter, the chapter is concluded.

IS studies and theories

IS/IT solutions are a combination of hardware, software, and telecommunications networks that people build and use to collect, create, and distribute useful data, typically in organisational settings (Valacich & Schneider, 2010). Organisations make use of IT solutions in order to function efficiently and effectively. However, it is used not only by organisations but also by communities to conveniently fulfil their needs, such as shopping and banking transactions (Campbell, 2019; Roethke et al., 2020;). More often than not, IS research focuses on the development, implementation, institutionalisation, support, and management of IS/IT solutions within context and perspective. Each and every study in IS has fundamental attributes that identify it, primarily technology, systems, process, and people.

IS research is a practical approach to professional enquiry in any social situation and it has increasingly become focal in recent years. In the last two decades, IS research has expanded significantly to a point of adopting and applying theories from sociology and other fields. Research is fairly considered a practical exercise due to the fact that events are studied in their natural settings, whether through qualitative, quantitative, or both methods (Koivisto & Hamari, 2019; Sarker et al., 2018).

Every research has objective(s); otherwise, there is no focus or forward-looking achievement. Thus, the objectives are the steps taken to answer the research questions. Objectives are specific statements that define measurable outcomes of research (Farrugia et al., 2010). In other words, the research objectives are the specification of the final reason for carrying out research. They help to develop a specific list of achievements that are required of the phenomenon being studied. Only thereafter can a theory be selected to underpin the study. Therefore, aims and objectives must be highly focused and feasible.

Until recently, 15 years to be precise, data analysis in IS studies was carried out using mainly the interpretative approach (Gregor, 2006; Walsham,

1995). This approach by itself is a challenge, as it does not have a formula and it is not guided by distinct views and boundaries. The argument could be attributed to the fact that people (researchers) often have different opinions and interpretations. Hence, it is useful and critically important to underpin a study with a theory or a combination of theories. Sociotechnical theories aim to underpin IS research, meaning that they provide guidance from the data collection stage to empirical data analysis. This has been extensively discussed in Chapters 4–9.

It is vital to consider the most appropriate theory to underpin the subject to be researched. Maschi (2016) argues that the theoretical framework can be thought of as a map or travel plan of the research. It serves as the structure and support for the basis of the research, the problem statement, the objectives, the significance, as well as the research questions (Grant & Osanloo, 2014). The theory that underpins the research is used as a lens to zoom into the data collected. It enables the ability to focus in addressing the research problem, research objectives, and questions, as well as the significance of the research. As a result, the findings are discussed in terms of how they relate to the theory that underpins the research.

Another significant aspect of reality that must be taken into consideration is the relationships that exist between the components that are involved in IS research. Therefore, without a relationship between the components, there is no interaction. This ultimately nullifies the existence of an IT solution. For example, there is interaction between software and hardware; otherwise, the computer-based system will not be able to execute its objectives or coded instructions. Computer hardware provides the underlying physical foundation for an organisation's IT infrastructure, and for storage and operation, it requires other infrastructural components, such as software, data, and networks.

The role of research methodology

The methods are briefly discussed because they are not the focus of the chapter. The intention is to show their association with types of data that determine the selection of different types of sociotechnical theories. The research methodology consists of methods, techniques, and approaches, which provide guidance to research in many fields, including IS. Zikmund (2003) portrays the criticality of research methodology as a master plan by explaining how it guides collecting, analysing, and interpreting data towards achieving the objectives of the research. In the same vein, Mouton (2011) emphasises that research methodology focuses on the research process and the kind of tools and procedures to be used in achieving the objectives of the study.

Research that is empirical or experimental in nature employs a class of methods, techniques, and approaches in the form of observations or interviews in collecting data in order to answer specific questions. Empirical research methods are often divided into two categories, namely, quantitative and qualitative methods. However, the methods can be combined as a mixed method. According to De Vos et al. (2011), quantitative and qualitative research methods are the two well-known and recognised approaches in IS research. The two methods are not in contrast; rather, they complement each another.

The qualitative study is concerned with non-statistical methods. It enables researchers to study social and cultural phenomena (Myers, 2009). The qualitative research method allows the studies to be carried out in their natural settings, attempting to make sense of or interpret phenomena in terms of the meanings people bring to them (Denzin & Lincoln, 2005). The interpretative nature of a qualitative research uncovers and describes the actual meanings that people associate with things in real-life settings. Also, interpreting is aimed at producing an understanding of the context of IS studies, whereby IS influences and is influenced by its context (Walsham, 1993). The qualitative approach takes social context into consideration in the construction, whether technical or cultural aspects, of an IS study. It acknowledges that, although IS has a strong technological component, its solutions are implemented and used by people within the social context.

The alternative to qualitative is the quantitative research method which is aimed at enabling researchers to also study natural phenomena to measure the social world objectively, as well as to predict and understand human behaviour. De Vos et al. (2011) describe the quantitative method as an enquiry into a social or human problem, based on the testing of a theory composed of variables, measured with numbers and analysed with statistical procedures in order to determine whether the predictive generalisations of the theory hold true. In quantitative studies, the interpretation of numbers is often seen as strong evidence of how and whether the phenomena worked or did not work. Based on this premise, it is argued that it is difficult to underpin numerical data with a theory. This chapter therefore focuses on the selection of theories that underpin qualitative studies.

The role of the research methodology is critical in selecting theories in that it defines the engineering components of a study through which the objectives are achieved. The methodology helps to define research attributes such as 'What' (strategy), 'Where' (design), and 'How' a study is carried out. The methodology starts with what type of data and materials are to be collected, which in turn influences the theory that is preferred in a research:

a What: The type of data and materials collected in order to achieve the objectives of the study is strategic in nature. The strategy is fulfilled through qualitative, quantitative, or both methods, which determines or influences the selection of theories.
b Where: This is the design of the study. It determines the entity to be studied, using methods such as the Case Study. This informs cultural settings and relationships within organisations.

c How: The 'How' attribute defines the techniques which are employed in data collection. Data is collected based on the strategy (What) and design (Where) as briefly described above. How the analysis or interpretation of the data is conducted remains a vital and critical stage of any study which draws on the theory that underpins the research.

d Why: On each of the research methodological steps that are employed in a study, there have to be justifications (Why). The justification helps to understand the flow and relationship amongst the methods, techniques, and approaches. Otherwise, it could be regarded as falsified or lacking basis. The justification helps to rationalise the selection of theories.

In the absence of a methodology, it is difficult or impossible to carry out studies which are rigorous, complex, or comprehensive in nature. Also, methodology helps to answer questions such as where, when, and how the study was conducted, as briefly described above. Another important factor on the roles of research methodology is the sequence in which the selected methods, techniques, and approaches are employed in a study. The importance of this factor is based on the fact that certain actions or events depend on one another. For example, the objectives of a study must be articulated before a decision on whether the study would be qualitative, quantitative, or both is reached. Similarly, the qualitative-ness of the study determines the theory that underpins the study. Otherwise, according to Jabar et al. (2009), researchers may not recognise the systematic approach that is provided by theory, in specific steps, during data analysis or interpretation.

Underpinning theory

The emergence of sociotechnical theories to underpin IS research has made research more interesting and or challenging, depending on individuals' perspective. The challenge has become very prevalent and it impacts the space of research activities in IS. This leads to a question that has now become popular amongst IS researchers, including among students: 'what theory can I make use of in my study?' To answer such a question, the choice of a theory to underpin an IS study is based on the research objectives and the type of data, primarily. As such, the objectives must be well articulated and understood. An understanding of the influences and drivers of the objectives is imperative. Also important is having a good understanding of the research problem.

The use of the theories to underpin studies is mainly in the area of qualitative interpretive research. This is primarily because of the interplay between technical and non-technical factors involved in the field of IS. Underpinning theories are referred to, by Gregor (2006), as theories for understanding the social context in IS studies. The theories are intended to explain 'how' and 'why' various factors connect with each other to produce and reproduce results that influence the deployment and management of IT solutions.

In IS research, sociotechnical theories are often applied as lenses. In Orlikowski (2000), the word 'lens' is used in the sense of assessment, where certain features are focused upon and emerge, and where the rest of the picture falls into the background. Lenses are used as analytical tools to aid the interpretation and analysis of data in research. The analysis of data determines and shapes the results of the study. Thus, analysis or interpretation of data is fundamental to IS research. Hence, the tool (theory) used in the analysis is deemed critical. How data is collected and analysed is within the frame of the theory which underpins the study. Therefore, the most appropriate theory or theories should be selected.

One of the significant factors of underpinning theories is that they encompass both technical and social contexts of the phenomenon under study. A theory which underpins a study is characteristically relied upon for rationales such as:

a To help establish the types of relationships; the dependence and influence of the relationship that exists between actors in the selection, development, implementation, and practice of IT solutions in an organisation.
b Provides guidance in the interpretation of empirical data gathered over time and within a context.
c Creates awareness of social events, processes, and activities which take place in the development, implementation, and practice of IT solutions.
d Reduces the gap of assumptions and predictions of actions within a context.

Applying sociotechnical theory to underpin a study can be viewed as the heart (core) of a research. In addition to the research objectives, the type of data helps to select sociotechnical theory. Figure 10.1 shows how the components connect with each other. It begins with qualitative, to determine the theory that can underpin a research and help to shape the result by virtue of examining the processes involved in the technology and non-technical artefacts in a phenomenon being studied.

As illustrated in Figure 10.1, qualitative research lends itself to narrative description. In this context, narrative is the process of organising the data in units. Roberts (2002) points out that in the process of data analysis or interpretation, it becomes evident that parts of the narrative are not drawn randomly but selected subjectively. The data is thereafter analysed following the interpretivist approach, using theory in order to gain deeper insights and comprehension of the relationship that exists amongst the stakeholders of an IT solution, which are often Process, Technical, and Non-technical.

The theory that underpins a study is potentially to help find the factors that make a difference in the development, implementation, practice, or at all stages of IT solution in organisations. Theories that underpin a study do not predict the outcome of the study. They are aimed to gain a better understanding of the sociotechnical dynamics in organisations and how people

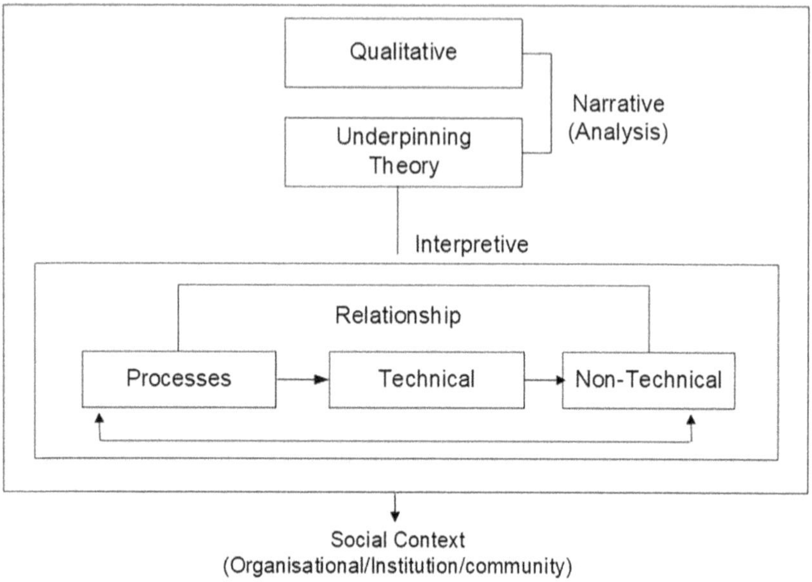

Figure 10.1 Research strategic approach.
Source: Author.

interact with IT solutions to solve current or emerging problems. Rather, they enhance our understanding of a situation, and through it an explanation is offered on why things happen in the way that they do. The theories make significant contributions to IS research through their theorisation of IS/IT solutions, particularly from the sociotechnical perspectives.

Selecting a sociotechnical theory for IS studies

Selecting a theory is not as easy or simple as it is sometimes claimed. Also, the know-how about a particular theory is not enough, or should not be used as the deterministic factor for selecting it. Rather, the choice of a suitable theory or theories should be based on the objectives of the research and should be influenced by the type of data. Nilsen (2015) suggests that a theory helps provide a clear explanation of how and why specific relationships lead to explicit events or actions.

Table 10.1 shows the most applied theories in IS research in the last two decades. This conclusion was reached, based on our analytical extraction from the google scholar search engine. The search was conducted on 11 sociotechnical theories that have been used in IS studies in the last two decades. These are ANT, AT, behavioural theory, cognitive theory, CT, DOI theory, grounded theory, narrative theory, organisational theory, ST, and TAM. The focus is therefore placed on the theories that have been applied in IS research over 50,000 times in peer-reviewed articles and books. As shown in Table 10.1, seven theories meet this frequency of use in IS studies.

Table 10.2 presents a brief description of the focus of each of the six theories. Each of the theories has been comprehensively covered in detail from Chapters 4 to 9. The selection and application of theories is influenced by various factors primarily because they constitute an integrated set of attributes. Also, theories are considered good if they have been tested or can easily lead or facilitate findings for generalisation.

Table 10.1 List of underpinning theories

Sociotechnical theory	Frequency of use in IS studies
AT	3,380,000
DOI	1,920,000
ANT	602,000
ST	107,000
CT	70,800
TAM	60,000

Table 10.2 Sociotechnical theories

Theory	Description
AT	AT primarily focuses on a structural property of events through which outcomes are reached. The main tenets of the theory include the subject (human actors), the object (things to be produced), tools (vehicle for producing things, such as technology), and the community (social system). In AT, human actors that form part of the social system make use of tools to create things. The social systems are formed and exist through and within rules. According to an African adage, a tree cannot make a forest. No one actor can produce a thing by itself alone. Hence, the division of labour within a community of actors in a social system is crucial (Barnard, 2010). Therefore, the human actors have the power to make a difference within social systems. Thus, AT is often applied as a lens to underpin studies, and to examine and better understand human activities within an environment (Hasan & Kazlauskas, 2014).
ANT	The ANT focuses on a constant shift of negotiation and relationship towards achieving goals and objectives. Actors are human and non-human. By implication, anything can be an actor in ANT as long as it has the power or ability to make a difference to a network. In ANT, both human and non-human actors are treated equally because it is argued that they offer equal contributions to the establishment and operations of a network (Chen et al., 2009). A network can be part of other networks, which makes actors and networks heterogeneous. According to Watson, Boudreau, and Chen (2010), power is not about an actor possessing it, but rather being able to use it to enrol, enlist, and convince other actors to follow them in a course that they define.

(*Continued*)

Theory	Description
CT	CT is a theory that claims that there is no specific best way to organise an organisation, or to make decisions (Volberda et al., 2012). Instead, some contingency factors, either internal or external, can influence an organisation's performance (Ganescu, 2012). Islam and Hu (2012) argue that organisational effectiveness is dependent on a fit between certain factors, which include the types of available technologies, environmental volatility, the size of an organisation, features of an organisational structure, and its IT solutions. Thus, performance is a process that interacts with the relationship between environment and strategy, and that in turn encourages organisational structure, for which IT solutions can be used to enable and support.
DOI	DOI seeks to explain how, why, and at what rate innovation (new ideas or technology) is diffused within a social system (organisation of people) (Jakki, Sanjit & Stanley, 2010). The theory is particularly concerned with how to take a new idea or technology to the community or people. However, no one person or group of people is forced to accept or reject the new idea or technology. Based on communication, interaction, and interpretation, the innovation is either accepted or rejected by agents (Sáenz-Royo, Gracia-Lázaro & Moreno, 2015). Whether the outcome is to accept or reject, the decision has to be implemented and confirmed by the stakeholders or relevant actors.
ST	ST focuses on agency and structure in a social system. Agent (or agents) can be technical and non-technical (such as human). Structure is the rules and resource in structuration (Iyamu & Roode, 2010). The theory puts emphasis on how events of social systems are produced and reproduced over time and space (Ma, 2010). In structuration, social systems exist because there are structures, and structures exist within social systems. Human agency and social structure are not two separate concepts or constructs but are two ways of considering social action termed duality of structure (Lamsal, 2012). The structure is the recurrent patterned arrangements which influence or limit the choices and opportunities available, whilst agency is the capacity of individuals to act independently and to make their own free choices.
TAM	TAM is an IS theory that models how users come to reject or accept and use a technology. The theory consists of two main beliefs, which are perceived usefulness (PU) and perceived ease of use (PEoU) that determine attitudes towards adopting new technologies (Erasmus, Rothmann & Van Eeden, 2015). The adopter's attitude towards new technology can be either positive or negative in an attempt to adopt the technology solution. In TAM, PEoU and PU are deterministic factors for individuals' or groups' decision for IS acceptance (Lee, Kozar & Larsen, 2003; Surendran, 2012). Erasmus et al. (2015) suggest that the rationale is that the influence of external variables on technology acceptance behaviour is mediated through user beliefs and attitudes, in which beliefs represent a degree of instrumentality tied to action, and attitudes are purely affective.

Based on the broad coverage and focus of the theories as shown in Table 10.2, it can realistically be challenging to select a theory to underpin an IS research. This can be attributed to the fact that many IS studies focus on development, implementation, management, and use of IS/IT, which involves interaction, relationship, and environment. Theories underpin studies with the primary intention to diffuse rigour and examine the sociotechnical contexts. Hence, the selection of an appropriate theory is critical. Otherwise, it could add complexity to the study. However, the selection of a theory or a combination of theories to underpin a study has never been easy. This can be attributed to many factors such as:

a Understanding – as mentioned earlier in the chapter, a majority of the theories which are used in the IS field originated from the sociology discipline. This makes it challenging for IS researchers to understand their contexts and focuses.
b Knowledge – some IS researchers are knowledgeable about a theory. This group of researchers attempts to make use of the particular theory in all the research types that they conduct.
c Application – many IS researchers (students included) face difficulties in applying sociotechnical theories in their studies. Experience has shown that even IS researchers who seem to understand some of the theories find the application challenging.
d Differentiation – some of the social theories have many similarities in their focus and coverage. For example, ANT and ST are similar with regard to the power relationship. In addition, ANT and ST focus on human and non-human actors, and technical and non-technical agents, respectively. The focal point of both theories is in negotiating interest within a system.

Table 10.3 highlights the tenets of the theories which can help in their possible selection to underpin IS research. The table consists of the most popular theories that are briefly described in Table 10.2. The description and tenets are considered as added influencing factors in the selection of the theories for IS studies.

An obvious fact is that the objectives and scope of studies differ in size; some are broader than others. In such instances, multiple theories might be required in order to gain full coverage of the data analysis. Also, the theories have the potential to help develop methodological frameworks which are based on a logical appropriation of its tenets and consistency (Avison et al., 2002). The tenets play a vital role in guiding the analysis of the data.

There should not be conflict or contradiction in the combined use of theories in IS studies. The aim is often not to compare or contrast the theories when they are applied complementarily in a study. According to Iyamu and

Table 10.3 Description and tenets

Theory	Description	Tenets
AT	Interpretative, narrative	Social context, agents, relationship, regulations, process-oriented
ANT	Iterative, interpretive, narrative	Power relation, varying interests, human, non-human, networks, process-oriented.
CT	Interpretative, narrative	Social context, process-oriented, cultural, human, intents
DOI	Interpretative, narrative	Social context, process-oriented
ST	Interpretive, narrative, multi-dimensional	Social context, power relationship, technical, non-technical, process-oriented
TAM	Interpretative, narrative	Assumption, presumption, human, predictive, technology, systems

Roode (2010), the aim is to highlight their importance and complementary usefulness in the research.

As shown in Figure 10.2, there are three fundamental components that are required to guide the selection of a theory or theories to underpin IS studies, which are: (i) the phenomena being studied; (ii) the objectives of the study; and (iii) the focus of a sociotechnical theory. These components influence and determine the selection of a theory or theories for IS research. This is mainly because the components are tightly connected and related to a point of irreversibility. This is primarily to avoid gaps in achieving the aim of the research. The connection or point of contact in a relationship can take many forms (Jones & Warren, 2016). With nonexistence of a relationship between and among the components, it is highly possible to have a disconnect, which creates a gap in the process. Reis, Collins, and Berscheid (2000) argue that relationships are about being and the experience of connecting between subjects and objects over a period of time. A relationship therefore constitutes and is enforced by connectivity, communication, and collaboration.

1. **IS studies**

   Each and every IS research has basic factors that are identified with them. Primarily, IT solutions involve three factors, technologies, people, and processes, as shown in Figure 10.2. The success or failure of an IT solution depends on how the factors are brought together in the selection, development, implementation, and practice. Therefore, IT ceases to exist in an organisation if any or a combination of the factors is missing or there is a disconnect. Hence, in many IS researches, the focus is primarily on investigating and understanding the roles of people, technology, and process, and what, why, and how they impact on IT solutions in organisations.

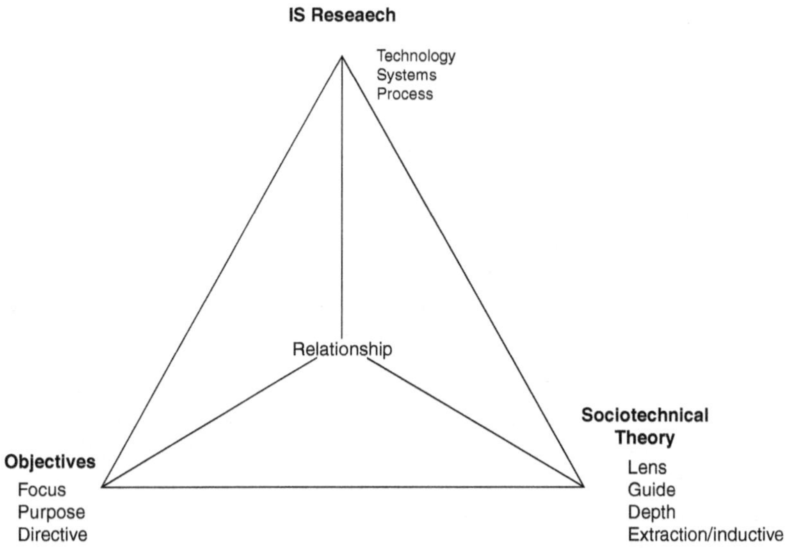

Figure 10.2 Theory selection components.
Source: Author.

a **Technology** – Technology is concerned with gaining insights and understanding how knowledge is applied with tasks involving people and machines to meet sustainable goals. Those who are involved in the technology require skills and knowledge to elucidate the rendering of products and service delivery. People need to follow various techniques, methods, as well as processes to arrive at the final product. Many organisations use technology to remain competitive. People create new products and services using technology and, through the products and services, create technology.

b **Systems** – Technologies are required to build systems. A system is a set of artefacts connected together to form a complete functioning product. In this context, a good example is an IS solution. In order to have a functional IS, it needs to be developed or selected from the shelf, based on requirements. It is a complementary network of hardware and software that people and organisations use to collect, filter, and process, create, and also distribute data. It enables organisations to function efficiently and effectively for competitiveness and sustainability purposes.

c **Process** – There are sets of rules and regulations that need to be adhered to in order to execute tasks at various stages in the development, implementation, support, and management of IT solutions in organisations. At the same time, the rules and regulations are either created or enabled and supported by technologies. In structuration, this is referred to as reproductivity of events and actions. Also, some of the processes are tools (such as software, hardware, and network protocols) or methodologies that must be employed, how they must be applied, and when they must be used. This implies that there is an inseparable connection and relationship between tools or methodologies and applying them in the delivering of IT solutions, which get empirically revealed in IS research. The connection and relationship are consciously or unconsciously enforced through interactions: human-to-human, human-to-technology, or technology-to-technology. Through such an understanding of connectivity, relationship, and interaction between actors, a sociotechnical theory can be selected to underpin the study.

The discussion above is intended to help researchers, particularly postgraduates, gain knowledge of how the three components connect or can connect to interrelate in the phenomenon that is being studied. Also, the components can be further decomposed into varying activities. The activities include the roles of the components in the problem that is being solved through the research. This understanding helps to align the components with the focuses and tenets of the different theories.

2 Objectives

The objectives of the research inform the methodological approaches that are employed in the study. This includes the development of realistic and relevant research questions and the selection of appropriate research design methods and analysis techniques. According to Farrugia et al. (2010), the objectives guide the focus and direction of each study. Thus, the objectives are specific, explicit, or predictive (hypothesis). It also helps to justify the execution and evaluation of a research. As a result, research objectives must be clear, concise, and declarative in statements, and must provide direction in investigating the various variables or elements that are connected with the phenomenon. Each of the variables or elements is further decomposed towards achieving the objectives.

An objective is an entity that draws on examining factors of influence. In IS research, the factors are often decomposed into technical (such as technologies) and non-technical (e.g., humans, rules, regulations, and the environment). No factor acts alone; it interconnects with or manifests from a relationship, and no relationship happens without interaction. These factors help a researcher to interrogate the phenomenon, using the what, how, where, and why interrogatives. Attempts to answer the questions guide the selection of theories and help to assess or evaluate why things are how they are in an environment.

3 Sociotechnical theory

As presented in Table 10.2, the focus and coverage of sociotechnical theories vary. The variables or variations allow explicitness and implicitness from the perspectives of rigour and expansion in IS research. The theories are employed to underpin IS studies, meaning that they are particularly to guide data analysis and interpretation of findings. Hence, they are applied as lenses to increase the depth of view in the analysis or interpretation of data. Theories give researchers different lenses through which to look at complex problems of a diverse nature, including social issues, and focus attention on different aspects of the objectives and data. Therefore, theory helps to guide researchers, from various angles of the research and its objectives.

4 **Relationship**

A relationship constitutes and is enforced by connectivity, communication, and collaboration. Consequently, without a relationship between and among an IS study, its objectives, and sociotechnical theory, there is a disconnect, which creates a gap in the process and affects the outcome of the study. Sociotechnical research is premised on the interdependent and inextricably linked relationships among the features of any technological object or system and the social norms, rules of use, and participation by a broad range of humans (Sawyer & Jarrahi, 2014).

i **Connectivity** – For a relationship to exist between actors (such as IS study, objectives, and theory), there must be a duality of interest endorsed by an interconnection. The basis of the duality lies in the relationship that the agency has with the structure. In the duality, the agency has much more influence on its lived environment than past structuralist theory had granted (Giddens, 1979). This type of connectivity is the link or allied interest that draws the components together. The interest and connection which people have manifest in the outcome or result of the objectives of the subject under study. Also, the interest and interconnection exist based on the how and what the actors communicate and interpret. Ghaffarian (2011) posits that the social-technical approaches can therefore complement mainstream IT research by capturing the contextual dynamics and actors' rationality, attending to relevant actors, their interests, and the logic of their negotiations or shifting interest.

ii **Communication** – Communication is a dual process of interaction that occurs in a relationship between actors, through which an interconnection is established. Through communication, intended meanings that are associated with things are conveyed between actors of the same entity or group or with another. The conveyance of the communication is enabled and supported, using technologies created by the same actors. According to Pettersson (2000: 64), there is 'a pattern that is always present where meaning is conveyed and understood between people'. But in order to do that, a theory must be used to guide or facilitate the processes of associated meaning, and the understanding and use of IT solutions in an environment, which are sometimes unique.

iii **Collaboration** – This is to ensure collectiveness and inclusivity, thereby avoiding isolation of certain factors or actors that might be of significant use. However, for this to happen, there must be a relationship which occurs through connectedness of actors within a social system. Through a 'connective' relationship, an organisation can also adjust its service offerings to exceed the customers' expectations (McCabe, 2010). This enables actions in the pursuit of objectives. For example, implementation of an IT solution involves collectiveness or collaboration between people, processes, and enabling technologies, which can be assessed or evaluated with a theory.

Conclusion

It is well documented that when embarking on IS research, there are various key factors that need to be taken into consideration. What has been missing is how to select a theory or theories to underpin IS studies. This chapter highlights the critical and fundamental areas that must be taken into consideration in the quest to select a theory for IS research. As proposed in the chapter, there are three main fundamental components that influence the selection of sociotechnical theories in IS research. The relationship between these three components contributes to the rigour and overall success of the research by achieving its aim and objectives. The next chapters dwell on complementarity of theories in IS research.

References

Avison, D. E., Wood-Harper, A. T., Vidgen, R., & Wood J. R. G. (2002). A further exploration into information systems development: The evolution of Multiview 2. *Information Technology and People, 11*(2), 124–139.

Barnard, R. (2010). Activity theory: A framework for analysing intercultural academic activity. *An International Journal of Human Activity Theory, 4*(1), 25–37.

Campbell, D. E. (2019). A relational build-up model of consumer intention to self-disclose personal information in e-commerce B2C relationships. *AIS Transactions on Human-Computer Interaction, 11*(1), 33–53.

Chen, W., Zhang, C., Zheng, Y., & Cui, L. (2009). The interpretive flexibility of an e-government project: From an actor-network theory perspective. *Proceedings of the 42nd Hawaii International Conference on System Sciences.*

De Vos, A. S., Delport, C. S. L., Fouche, C., & Strydom, H. (2011). *Research at grass roots: A primer for the social science and human professions.* Pretoria: Van Schaik Publishers.

Denzin, N. K., & Lincoln, Y. S. (2005). *The SAGE handbook of qualitative research* (3rd ed.). Thousand Oaks, CA: Sage Publications.

Doolin, B. (1998). Information technology as disciplinary technology: being critical in interpretive research on information systems. *Journal of Information Technology, 13*(4), 301–311.

Erasmus, E., Rothmann, S., & Van Eeden, C. (2015). A structural model of technology acceptance. *SA Journal of Industrial Psychology*, *41*(1), 1–12.

Farrugia, P., Petrisor, B. A., Farrokhyar, F., & Bhandari, M. (2010). Research questions, hypotheses and objectives. *Association Médicale Canadienne*, *53*(4), 278–281.

Ganescu, M. C. (2012). Assessing corporate social performance from a contingency theory perspective. *Procedia Economics and Finance*, *3*, 999–1004.

Ghaffarian, V. (2011). The new stream of socio-technical approach and main stream information systems research. *Procedia Computer Science*, *3*, 1499–1511.

Giddens, A. (1979). *Central problems in social theory*. London: The Macmillan Press Ltd.

Grant, C. & Osanloo, A. (2014). Understanding, selecting, and integrating a theoretical framework in dissertation research: Creating the blueprint for your "house". *Administrative Issues Journal: Education, Practice, and Research*, *4*(2), 12–26.

Gregor, S. (2006). The nature of theory in information systems. *MIS Quarterly*, *30*(3), 611–642.

Hasan, H., & Kazlauskas, A. (2014). Activity Theory: Who is doing what, why and how. In: H. Hasan (eds.), *Being practical with theory: A window into business research*, pp. 9–14. Wollongong, Australia: THEORI.

Islam, J., & Hu, H. (2012). A review of literature on contingency theory in managerial accounting. *African Journal of Business Management*, *6*(15), 5159–5164.

Iyamu, T., & Roode, D. (2010). The use of structuration theory and actor network theory for analysis: Case study of. *Social Influences on Information and Communication Technology Innovations*, *2*(1), 1–17.

Jabar, M. A., Sidi, F., Selamat, M. H., Ghani, A. A. A., & Ibrahim, H. (2009). An investigation into methods and concepts of qualitative research in information system research. *Computer and Information Science*, *2*(4), 47–54.

Jakki, M., Sanjit, S. & Stanley, S. (2010). *Marketing of high-technology products and innovations* (3rd ed.). New York: Pearson.

Jones, P., & Warren, S. (2016). *Creative economies, creative communities: Rethinking place, policy and practice*. London: Routledge.

Koivisto, J., & Hamari, J. (2019). The rise of motivational information systems: A review of gamification research. *International Journal of Information Management*, *45*, 191–210.

Lamsal, M. (2012). The structuration approach of Anthony Giddens. *Himalayan Journal of Sociology & Anthropology*, *5*(4), 111–122.

Lee, Y., Kozar, K. A., & Larsen, K. R. T. (2003). The technology acceptance model: Past, present, and future. *Communications of the Association for Information Systems*, *12*(1), 752–780.

Lim, S., Saldanha, T. J. V., Malladi, S., & Melville, N. P. (2013). Theories used in information systems research: Insights from complex networks analysis. *Journal of Information Technology Theory and Applications*, *14*(2), 5–44.

Ma, L. (2010). *Theory and education: A case of structuration theory*. In Proceedings of iConference, Urbana-Champaign, Illinois, 3–6 February.

Maschi, M. B. (2016). *The case for explicit, skills-based resilience building instruction*. University of New England.

McCabe, S. (2010). *Marketing communications in tourism and hospitality: Concepts, strategies and cases*. Amsterdam: Elsevier.

Mouton, J. (2011). *How to succeed in your master's and doctoral studies: A South African*, p. 280. Pretoria: Van Schaik Publishers.

Myers, M. D. (2009). *Qualitative research in business & management.* London: Sage Publications.

Nilsen, P. (2015). Making sense of implementation theories, models and frameworks. *Implementation Science*, *10*(1), 1–13.

Orlikowski, W. J. (2000). Using technology and constituting structures: A practice lens for studying technology in organizations. *Organization Science*, *11*(4), 404–428.

Pettersson, A. (2000). *Verbal art: A philosophy of literature and literary experience.* Montreal: McGill-Queen's Press-MQUP.

Reis, H. T., Collins, W. A., & Berscheid, E. (2000). The relationship context of human behavior and development. *Psychological Bulletin*, *126*(6), 844–872.

Roberts, A. (2002). A principled complementarity of method: In defence of methodological eclecticism and the qualitative-quantitative debate. *The Qualitative Report*, *7*(3), 1–18.

Roethke, K., Klumpe, J., Adam, M., & Benlian, A. (2020). Social influence tactics in e-commerce onboarding: The role of social proof and reciprocity in affecting user registrations. *Decision Support Systems*, *131*, 113268.

Sáenz-Royo, C., Gracia-Lázaro, C., & Moreno, Y. (2015). The role of the organization structure in the diffusion of innovations. *PLoSONE*, *10*(5), e0126076.

Sarker, S., Xiao, X., Beaulieu, T., & Lee, A. S. (2018). Learning from first-generation qualitative approaches in the IS discipline: An evolutionary view and some implications for authors and evaluators (PART 1/2). *Journal of the Association for Information Systems*, *19*(8), 752–774.

Sawyer, S., & Jarrahi, M. H. (2014). Sociotechnical approaches to the study of information systems. In: H. Topi & A. Tucker (eds.), *Computing handbook, third edition: Information systems and information technology*, pp. 5–10. New York: Chapman and Hall/CRC.

Surendran, P. (2012). Technology acceptance model: A survey of literature. *International Journal of Business and Social Research*, *2*(4), 175–178.

Valacich, J., & Schneider, C. (2010). *Excerpted from information systems today: Managing in the digital world* (4th ed.). London: Prentice-Hall.

Volberda, H. W., van der Weerdt, N., Verwaal, E., Stienstra, M., & Verdu, A. J. (2012). Contingency fit, institutional fit, and firm performance: A Metafit approach to organization–environment relationships. *Organization Science*, *23*(4), 1040–1054.

Walsham, G. (1993). *Interpreting information systems in organizations* (Vol. 19). Chichester: Wiley.

Walsham, G. (1995). Interpretive case studies in IS research: Nature and method. *European Journal of Information Systems*, *4*(2), 74–81.

Watson, W. T., Boudreau, M., & Chen, A. J. (2010). Information systems and environmentally sustainable development: Energy informatics and new directions for the IS community. *MIS Quarterly*, *34*(1), 23–38.

Zikmund, W. G. (2003). *Business research methods* (7th ed.). Ohio: Thomson.

11 Activity theory and actor-network theory

Table of contents

Introduction

Many organisations continue to encounter various challenges at different stages of selection, development, implementation, innovation, support, and maintenance of information systems and technologies (IS/IT) solutions (Yujie & Xindi, 2010). Hinkelmann and Pasquini (2014) suggest that the challenges result from both technical and non-technical factors. Thus, information systems (IS) studies are regularly conducted to gain an understanding of how these factors manifest themselves to have a negative impact on IT solutions in organisations. Many of the studies focus on assessing and evaluating, to improve efficiency and effectiveness while exploring the trajectories of the challenges in IT solutions. Research is carried out in the IS field, primarily to understand what and why things happen, including the influences and impacts that they have on the environments where they are deployed. In their quest to achieve such aims, some IS researchers employ theories as a means to guide the analysis or interpretation of the data, in order to improve comprehension of the deployment, management, and use of IT solutions.

Chapters 4 and 5, respectively, covered activity theory (AT) and actor-network theory (ANT) in detail. As explained in the chapters, both theories are increasingly but separately applied in IS research to guide data analysis and interpretation beyond the norm (Iyamu, 2013). Both AT and ANT

DOI: 10.4324/9781003184119-11

draw on social and technical contexts, but from different angles. The vast and diverse scopes of IS/IT operations make it essential to apply multiple theories to underpin IS studies. This is primarily to ensure a comprehensive coverage of the diverse objectives and to avoid gaps in the studies. This necessitates complementarity of theories as presented in this and the chapter that follows. Atkinson and Brooks (2003) also complementarily employed structuration theory (ST) and ANT in their study. The theories of AT and ANT have been applied in IS studies over the years, but separately (Hasan, Smith & Finnegan, 2017; Wong, 2016). This chapter proposes applying AT and ANT complimentarily as lenses in IS research to guide the analysis and interpretation of data.

AT is a sociotechnical theory that is concerned with the development of social activities (Shaanika & Iyamu, 2015). Dennehy and Conboy (2017) suggest that the theory has been used as a conceptual framework for analysing complex and human activities within social systems. This is mainly because activity is a sum of actions that living beings carry out in order to achieve their goals, as done in IS research. An activity is a system of actions in pursuit of some socially determined object (Hasan et al., 2017). In the context, an object is the rationale or motive why an activity is initiated. To achieve these objectives, humans wholly depend on the technical tools and artefacts (tools) as a mediating means between them and the object. Tools can take various forms, depending on the context of the study; as such, they may range from instruments, signs, procedures, machines, language, methods, laws, or forms of work organisation (Er, Kay & Lawrence, 2010).

In the process of carrying out activities, the subjects that are involved consciously plan their actions and interact with other humans within the communities. Within each social system (community), there are rules that are implicitly or explicitly used to govern activities, as well as manage the interactions that happen in the process of allocating tasks and responsibilities among the actors (Karanasios & Allen, 2013). However, AT does not cover areas such as how interaction came into being, how an activity is problematised, and the consequence of relationship between humans. ANT focuses on these essential factors. Some important factors are the relationship and interaction between human and non-human actors in the development and implementation of IS/IT solutions in the various environments. In ANT, actors' interactions create associations and define one another through the intermediaries that pass between them (Callon, 1981).

ANT is considered a highly influential theory within the sociology of science that seeks to explicate and interpret sociotechnological development and change. The main tenets of ANT include actor and network (Baiocchi, Graizbord & Rodríguez-Muñiz, 2013). In ANT, actors are both human and non-human entities that have the capability of changing a certain state of affairs (Müller & Schurr, 2016). ANT is thus known for advocating a flat ontology between humans and non-humans in an environment, which is called agnosticism, as discussed in chapter 5. This refers to examining human and non-human actors equally, without priority being given to either technology or social issues about them (Wong, 2016). People, technology, and process are thus regarded to be of equal importance and value in ANT. Irrespective of whether the actor is human or non-human, they are both weighed equally and they are considered to offer the same contribution to a network (Wernick, Hall & Nehaniv, 2008).

The separate use of AT and ANT in some IS studies can possibly affect the results due to their scopes. The complementary use of both theories means bridging the gap that each theory cannot comprehend, if applied alone. Based on a study's focuses, a single theory may not be able to fully cover the objectives, hence the complementary use of theories. In Atkinson and Brooks (2003), ST and ANT were combined, not only to understand how human actors and facilities and their networks come into being, persist, and change, but also to envision and develop them. Iyamu and Roode (2010) highlight the importance of combining ST and ANT and emphasise the usefulness of complementarity of theories in IS studies, including an interrogation of the relationship between individuals and technology. There was no evidence on complementary use of AT and ANT in IS studies as of the time of writing this book.

The main question this chapter attempts is, How can AT and ANT be complementarily employed as lenses in IS studies? Hence, the focus is: (i) to investigate how both AT and ANT can be combined in IS studies, which includes identification of strengths and weaknesses of the theories; and (ii) to examine the order-of-use in a combined state. The remainder of this chapter is divided into four main sections. The first section is a review of literature in the areas of IS, AT, and ANT. The meeting points between AT and ANT, from the perspective of IS studies, are problematised in the second section, which highlights the strengths and weaknesses of both theories. The third section presents and discusses the framework which can be used to guide the complementarity of AT and ANT in IS studies. Finally, a conclusion is drawn.

Literature review

Based on the nature of the chapter, the literature review is conducted from the angle of three subjects: the field of IS, AT, and ANT from IS perspectives. In conclusion of the review, I summarise the application of the theories and IS studies.

Information systems

The field of IS constitutes and is recognised as the study of complementary software and information technology (IT) solutions, including network protocols, that people and organisations use to collect, filter, processes, create, and distribute data. Based on their coverage, IT solutions are often used to enable and support organisations in executing operations and strategies towards achieving their goals and objectives. However, IS/IT artefacts or solutions are characterised by constant change and subsequent developments and implementations. The change and development are driven by business demands for improved efficiency and effectiveness. Métrailler and Estier (2014) stated that in order to cope with change and new market opportunities, IS that supports these processes should be able to evolve in adequate and responsive ways.

Over the years, academics have through research responded reciprocally and empirically to the rapid developments and evolutions in the field of IS (Peppard & Ward, 2004). One of the efforts and responses is the innovative application of sociotechnical theories to underpin IS studies (Cordella, 2010). This includes gaining deeper insights into complex settings as challenges persist in the computing environments in many organisations (Shaanika & Iyamu, 2015). In general, sociotechnical theories are purposely employed in IS studies to examine, assess, or evaluate, so as to empirically elucidate the factors that inform the growing complex situations in IT. More often than not, IT solutions are of a complex sociotechnical nature built into systems that are implicitly and explicitly connected with people, other technologies, services, and processes that work together for the realisation of organisational goals and objectives (Karanasios & Allen, 2013). Varga (2003) defines IT as a composition of complex automated systems that cover the informational tasks of an organisation. An (2010) clarifies that IS/IT refers to a system of people, data records, and activities that process data and information in an organisation.

Based on scope and objectives, researchers in the field of IS employ sociotechnical theories to guide and underpin their studies. The theories are applied in IS studies with particular cognisance of and focus on three main components. These are people, processes, and technologies (Shim & Shin, 2016). Also, the theories draw on the interconnection among the components that are required to be further interrogated, assessed, or evaluated in order to better fathom how IS/IT solutions are deployed, used, and managed in organisations. Rose and Scheepers (2001: 217) state that 'social theory has a substantial part to play in the development of the discipline of IS, primarily by helping to understand and interact with the societal, organisational and personal contexts without which the technology is meaningless'.

Activity theory in Information Systems

As established in Chapter 4, AT is a social theory of human consciousness, interpreting consciousness as the product of an individual's interactions with people and artefacts in the context of everyday practical activity (Hasan et al., 2017). Moawad, Liu and El-Helly (2013) adduce that before actions are performed, they are planned in a conscious manner within a model. The model acts as a methodology for how the actions are carried

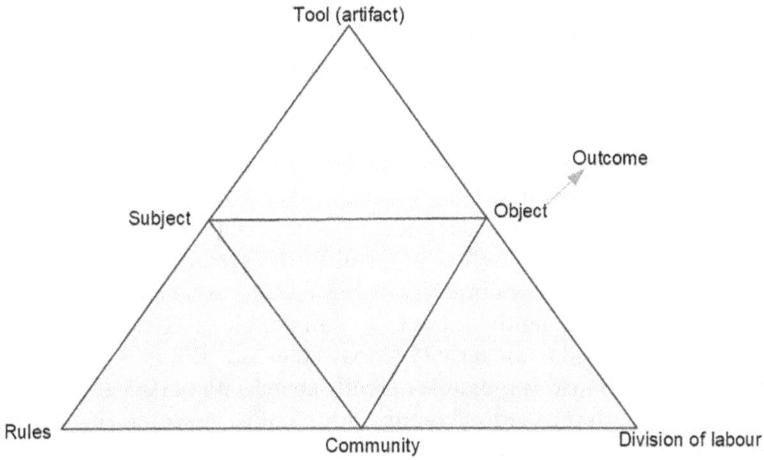

Figure 11.1 Activity theory.
Source: Engeström, Miettinen, and Punamäki (1999).

out, place, time, and impact. Even though the activities are developed consciously, their outcomes are not always as intended by the subject.

Figure 11.1 depicts the main components of the AT, which are tool, subject, object, rule, division of labour, community, and outcome (Tobin, Milne & Plakitsi, 2013). The interconnections and relationships between the components depend and draw on a comprehension of the entire activity system. The subject is thus the source of the activity and the activity is directed to the object (Dennehy & Conboy, 2017). In the view of ER et al. (2010), the subject undertakes an activity in order to achieve an object. In AT, a subject is an actor, a living being with specific needs.

Humans as subjects in an activity system are therefore considered superior and have the power to make a difference. Rose and Scheepers (2001) assert that power involves the exploitation of resources within a social system. The availability of resources influences the probability of attaining and defining an object. An object may be something materialistic, or at least tangible, such as a plan or an idea (Moawad et al., 2013). In addition, AT can be described as an entity that holds the connection between the components, as shown in Figure 11.1, where, for example, subjects use tools to carry out their activities to produce an object. The relationship between the subject and object is always mediated by one or more tools which can be instruments, signs, procedures, machines, methods, laws, or ways for the organisation of work (Moawad et al., 2013). Er et al. (2010) explained that the tools that humans make use of in carrying out their actions have an impact on the results of an activity.

In addition, tools assist or enable individuals or groups in the decisions that they make in the course of reproduction through division of labour (Karanasios & Allen, 2013). In AT, division of labour explicitly reveals individual roles, responsibilities, as well as accountabilities of a subject that is involved in an activity. Also, division of labour involves the division of power and status of individuals or groups within a community (Tobin et al., 2013). A community represents the social system within which subjects carry out their activities and interact with each other. Jonassen and Land (2012) define community as an interdependent system because of the collaborative efforts of its members, as well as in terms of the greater societal systems in which it is nested. In a community, subjects are governed by implicit and explicit rules. Consequently, these rules can enable or constrain the events or processes involved in an activity (Moawad et al., 2013).

Also, in AT, it is near impossible to study an activity outside its natural environment, such that its entire compositions can be captured (Kaptelinin & Nardi, 2006). Explicably, activities are not static. They evolve in a non-linear way. Hence, each activity has its own characteristics and history (Tobin et al., 2013). The changes occurring bring development to the activity. In an IS study, Shaanika and Iyamu (2015) explained how the development of an activity can have an impact on the associated components (tool, subject, rules, community, division of labour, and object).

Based on its connectedness to IS/IT solutions that are embedded in an organisational episode, AT is increasingly applied in IS research in various ways to assess, examine, or evaluate. This includes the study of knowledge elicitation that is essential in the development, implementation, and management of IT solutions in organisations, as well as the human-computer interaction (HCI). AT is applied in IS studies primarily because IS/IT consist of sociotechnical objects that are created by humans (objects), using different skills, through various processes (rules), for environments (communities). The activities carried out by humans as they develop, implement, maintain, and interact with the systems are complex in nature. The social context in IS research makes it more challenging, whether in the development or in the implementation stages. From the numerous studies where AT has been applied as a lens, it is clear that the theory does not only add rigour to the studies, but it also helps to unpack and understand complexities within IT solutions in an environment.

Centred on AT focus on actors, Kaptelinin and Nardi (2006) put perspectives together as follows: (i) an emphasis on human intentionality; (ii) the asymmetry of people and things; (iii) the importance of human development; and (iv) the idea of culture and society, which shape human activities and are the main tenets that have drawn AT to the study of people and the use of technology within different environments. However, the authors were also specific that, in AT, a subject has needs that can be met only by living beings and by acting in the world. Therefore, non-living entities, such as computers, can be regarded as objects. Significantly, this means that there will always be a lack of relationship between humans and non-humans. Therefore, in an IS study where human and non-human actors have a relationship and interaction, there is likely to be a gap. ANT fills the gap, in that among other things the theory is concerned with the relationship and interaction that exist between human and non-human actors in a network.

Actor-network theory in Information Systems

ANT originates from the field of sociology but it has for many years been adopted in other fields to analyse and interpret data in different types of studies (Baiocchi et al., 2013; Shim & Shin, 2016). The focus of ANT is on shifting negotiation and the associations that are formed between actors, resulting in networks and their transformation (Prado & Baranauskas, 2012). This is a state which Macome (2008) describes as an integration of human and non-human actors to create a stable network. Subsequently, the main purpose of having a network is to allow actors from various areas to collaborate and to solve a problem or create something new or different. According to Latour (1987), ANT addresses the elements of science and technology in the making, as opposed to ready-made science and technology. ANT's approach is thus to unpack and analyse the elementary components, such as actors, processes, power, environment, policies, and networks, which are often involved in the development, implementation, and management of IS/IT solutions. The moments of translation, from the perspective of ANT, is often employed as a lens in IS-related studies (Müller & Schurr, 2016).

Moments of translation

The moments of translation is a process where an activity transforms or is given a new meaning from one stage to another. In Prado and Baranauskas (2012), the process of building associations among actors is called translation, in which there is the desire to change a certain state of affairs, stimulate actors' interests to join, and define roles to ensure compliance with the responsibilities that are assumed. Translation is the mechanism by which the network builder recruits actors, capably allocates tasks, and ensures their faithful associations.

However, translation is not deterministic, for what actors do when they come together in a network is unpredictable (Fenwick & Edwards, 2010). Cordella (2010) clarifies that actors are influenced by the network and, in turn, they influence the network by negotiating their forces through translation. As shown in Figure 11.2, Callon (1986) demonstrates that translations of an activity by an actor or actors within a network are achieved through four stages referred to as 'Moments of translation', Problematisation, Interessement, Enrolment, and Mobilisation, as shown in Figure 11.2.

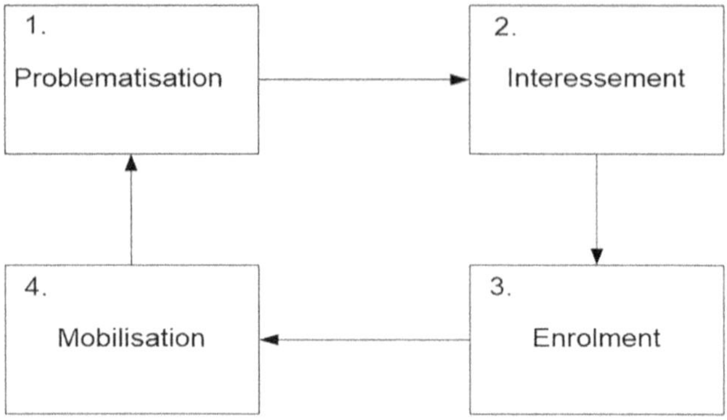

Figure 11.2 Four moments of translation.
Source: Callon (1986).

i *Problematisation:* it is the first stage of translation that deals with the identification of a problem, which is intended to create a solution. According to Callon (1984), problematisation is the process wherein a focal actor defines a problem in a manner that enables other actors to recognise it as their own problem, convincing them to join the network to provide a solution.

ii *Interessement:* it is the second stage, wherein interested actors join a network for a problem that has been problematised. Interessement is a series of processes by which the focal actors seek to lock other actors into the roles proposed for them in the programme.

iii *Enrolment:* at this stage, some interested actors within a network begin to participate in finding a solution to the problem. According to Macome (2008), only actors who accept the roles defined for them in the previous stage participate in building the network.

iv *Mobilisation:* it is the last stage of translation, meaning that all actors have been enrolled successfully into the network. When actors become connected, the consequences of success or failure spread through the network, so that there is mutual interest that everyone be successful (Prado & Baranauskas, 2012). It is also at this stage that a 'spokesperson' who can successfully translate the interests, roles, and relations of the entire network is able to predictably speak on behalf of the others (Callon, 1984). This stage translates into the successful establishment of an actor network with aligned interests.

Consequently, translation is a process that is undertaken by actors for a network to evolve and stabilise. Andrade and Urquhar (2010) claim that using ANT as a lens, the translation process enables them to exhume the trajectory of events. Exhuming events through the different stages of translation enhances the understanding of how networks came into being.

In practice, ANT is applied in IS studies to try to trace and explain the processes whereby relatively stable networks of aligned interests are created and maintained over a period of time, and to examine why such networks establish or fail to establish themselves (Walsham, 1997). It is within this context that Gao (2005) applied ANT to analyse qualitative data towards a strategy formulation process, through examining the components and elements that together form the social and technological viewpoints, and how they are dynamically connected within the environment. From a different perspective, AT is applied to examine and understand an existing activity (or action) towards its outcome rather than a collective of activity (Mursu et al., 2007). It, therefore, can be employed to assess the roles of rules and community, including individuals' responsibilities, but without venturing into relationship and power effect. These are areas ANT helps to bridge to unveil human and non-human power relations and how the interaction manifests to influence factors within a system.

Problematising complementary use of AT and ANT in IS studies

Sociotechnical theories are employed as lenses to underpin IS studies, which are intended to assist and intensify interrogation of data in the phenomenon that is being studied. Also, the theories help to amplify or 'dig deeper' into entities by using interrogative such as what, how, and why in the questions that are used to obtain responses about IT solutions in the computing environments of organisations. Depending on aim, either AT or ANT (as with other theories) may not necessarily be enough to underpin a study towards achieving its entire aim and objectives. Thus, AT and ANT can be used complimentarily in IS studies because of the uniqueness each of the theories brings to the fore. Both AT and ANT are comprehensively covered in Chapters 4 and 5, respectively.

However, there exist challenges in attempts to complementarily apply both theories in an IS research. Fundamentally, two main challenges exist: (1) rationalisation of order-of-use – which theory should be applied first in macro and micro analysis of data; and (2) understanding where either of the theories starts and ends. These challenges can be approached by knowing the tenets of each theory and understanding their strengths and weaknesses. The gaps or shortcomings in one of the theories can be filled by the other. This means that the use of both theories can be applied successfully, as proposed in this chapter.

AT focuses on activity. However, the theory lacks the process of establishing and problematising an activity. This is a gap ANT fills through the moments of translation in IS studies, as it focuses on how an activity is transformed from one stage to another by first acknowledging the problem – a process referred to as problematisation (Heeks & Stanforth, 2007). This process is followed by recruiting actors affected or connected (known as stakeholders) by the said problem. Despite its strengths, ANT has been criticised for its claim that both human and non-human actors are equal, meaning that the actors are viewed with equal importance and value within a network. The approach of ANT equating humans to the interaction of non-humans is contested (Cresswell, Worth & Sheikh, 2011). The justification for such criticism of dehumanisation is that it will be impossible for humans to create networks without non-human actors; thus, non-human actors also have the same power as the human actors. This and other criticism are exhaustively covered in chapter 5. This shortcoming of ANT is addressed by AT, where human actors and tools are distinguished, the former being superior to the latter. In AT, actors are subjects which are living things with needs to be fulfilled, thus their desire to execute certain activities (Shaanika & Iyamu, 2015).

In essence, even though both theories try to interrogate the establishment of IS/IT development and implementation, their main objectives differ. AT aims to interrogate the interaction between subjects within communities as they carry out their roles, using tools in accordance with set rules towards achieving objectives within a scope. Sannino, Daniels, and Gutiérrez (2009) give details on how AT is a theory of structure and content for specific output. Whilst ANT is concerned about how actors operate within heterogeneous networks through translation, ANT takes cognisance of the fact that multiple stakeholders could be involved in IS/IT development and implementation, creating different networks, thus trying to uncover all possible networks that are involved. It was affirmed by Walsham and Sahay (1999) that ANT examines the motivations and actions of actors who form elements, linked by associations of heterogeneous networks of aligned interests. ANT provides the interrogation of involved networks, which AT lacks.

Guide on the complementary use of AT and ANT in IS studies

Based on the sections above, a guide for complementary use of both AT and ANT in IS studies is developed. The guide is divided into two phases. This helps to achieve the objectives of the chapter as stated in the introduction section: (i) complementary use of AT and ANT in an IS study, illustrated in Table 11.1; and (ii) order-of-use of both theories in a study, as depicted in the framework (Figure 11.3). Each part of the guide can neither be separated nor used in isolation if both theories are to be successfully used complementarily in an IS study. The guide can be applied for data analysis or interpretation in the two phases, as follows:

i In the first phase, the objects and subjects in the phenomenon being studied are abstracted from both AT and ANT perspectives. This is done in a complemented fashion, as explained below and shown in Table 11.1.

Table 11.1 Complementary use of AT and ANT

Y *X*	*Object*	*Subject*	*Tools*	*Rules*	*Division of labour*	*Community*
Problematisation	What objects are problematised into activities within networks?	Who are the subjects that ensure and legitimately problematise the activity?	Why are certain tools used to problematise activities of networks?	How are rules used to legitimately problematise activities?	When and how are tasks, roles, and responsibilities of activities problematised?	Where is the community during problematisation of an activity?
Interessement	What are the factors that influence interest in objects in an activity?	Who are the subjects that are interested and willing to join in an activity?	Why are certain tools used to influence interest in an activity?	How are rules used to gain actors' interest in the activities of networks?	When and how are roles and responsibilities used to enact interest in activities?	Where do actors' interest in an activity spin from within the network?
Enrolment	What objects are used to participate in the activities of networks?	Who are the subjects that participate in the activities of networks?	Why are tools used to participate in activities of networks?	How are rules used to enforce actors' participation in activities?	When do actors accept the roles that are defined for them?	Where (at what point) do actors participate in activities of networks?
Mobilisation	What objects are used in mobilising other actors for activities?	Who are the subjects that mobilise other actors in the activities of networks?	Why are tools used to mobilise an actor within the activities?	How are rules used to mobilise other actors in an activity?	When is mobilisation carried out in an activity?	Where does mobilisation of activities occur?

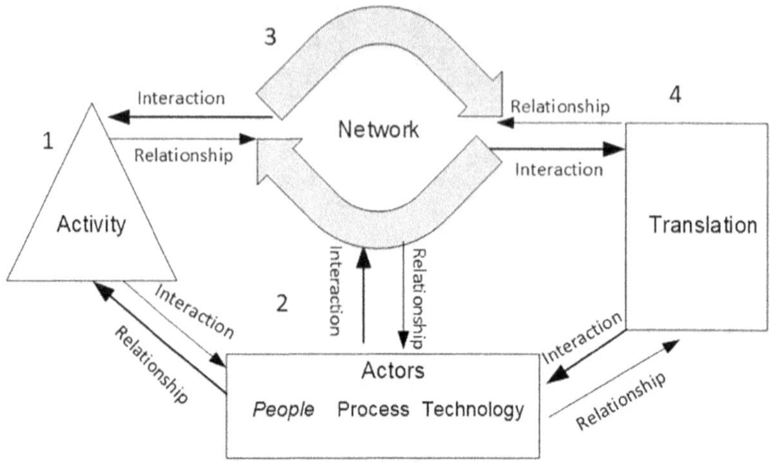

Figure 11.3 Order-of-use.
Source: Author.

ii In the second phase, the order-of-use is employed to identity and examine the activities, actors, and networks that exist in the development, implementation, management, and use of IS/IT artefacts or solutions. Also, the translation of activities, interactions, and relationships that happens between actors and within networks can be assessed or examined.

Complementary use of the theories

Table 11.1 represents a two-dimensional classification intended to guide researchers on how to complementarily apply the theories of AT and ANT, and to guide analysis or interpretation of data in an IS research. The six components of AT are presented in the '*X*' horizontal axis, while ANT's four moments of translation are vertically positioned on the '*Y*' axis. The prefixes of what, who, why, how, when, and where are used as the interrelationship and interconnection between the horizontal and vertical axes of the table. The cells in the table define the interrogative relationship that AT and ANT can be applied to share during analysis or interpretation of data. This approach fosters complementarity in applying the two theories in an IS research. Also, the intersections between the theories clarify any potential confusion and eliminate duplication in the course of applying the theories in a study.

The table provides a methodological approach through which information can richly be gathered from participants in qualitative studies. The approach is methodological in that it guides the researcher on a step-by-step basis, in accordance with the rows and columns of the table. This allows each question to be posed from two different, *Y* and *X* axes, viewpoints, towards exhaustive elaboration of views and opinions that subjectivism requires in qualitative studies.

The intersections are used to complementarily instil critical thinking and elicit deeper responses from participants. These are fundamentally required in IS studies, particularly qualitative research that is founded on subjectivism. It thus helps to gather rich data to gain a better understanding of the factors that influence, through enablement or constraints, the development, implementation, management, and use of IS/IT solutions in an organisation. Also, the intersection enables researchers to fill gaps any of the theories had left in vacuity and to enhance their strengths. Thus, the interaction of the components provokes researchers' thinking and understanding needed to establish the shortcomings of each component.

Order-of-use

The framework, as shown in Figure 11.3, is developed to guide complementary use of both AT and ANT in IS research. This is primarily to define the order-of-use which enables and distinguishes between the two theories in their complementary use in IS research. The framework clarifies the boundaries where one of the theories stops and the other begins in an analysis or interpretation of data. The main components of the framework include activity, network, translation, and actor. These are interconnected and depend on each other through the interaction and relationship that exist or are created in the process of selection, development, and implementation of IS/IT solutions. The discussion that follows should be read with the framework in order to gain a better understanding of the order-of-use.

Applying the framework that is shown in Figure 11.3 involves four main sequential steps:

- *Step One:* Identify the activities that are involved in the phenomenon being studied.
- *Step Two:* Identify the actors (people, process, or technology) and their roles in the development or execution of the various activities.
- *Step Three:* Identify the networks in accordance with the activities that are created, and the participating actors.
- *Step Four:* Translate the activities as they happen within the individual network. This is done through the interaction manifesting into negotiation, as well as the relationship that exists.

Activity

In the development, implementation, management, or governance of IS/IT solutions, there are many activities, including relationships and interactions that exist among involved humans and non-humans (Dennehy & Conboy, 2017). The relationship or interaction helps to examine, assess, or evaluate how and why people, processes, and technologies are interconnected to a point of inseparability within a network. From one perspective, the interaction defines and is defined by the relationship that exists within a network. From another angle, the relationship that exists and the interaction that happens among actors are the determining factors of influence and have an impact on IS/IT solutions, whether in development or implementation. Er et al. (2010) identified and explained that technical, social, and environmental factors are the main considerations that influence the deployment of technologies. The interaction among actors happens consciously or unconsciously in the process of carrying out activities of allied interest, which results in the establishment of relationships and can lead to stability.

Through action, interaction, and the relationship that transpires within a network, an activity is translated at different stages. The AT can be used to understand the associated components and intervention into the complex aspects of IS activities (Hasan et al., 2017). The translation can occur between actors in three ways (i) human-to-human; (ii) human-to-non-human; and (iii) non-human-to-non-human. It therefore clearly makes sense that an activity must exist before a network is identified or formed, within which translation of events takes place. Thus, neither translation nor network is established before an activity is created. As well documented and established in this book, AT primarily focuses on activity. It, therefore, should come first in the order-of-use, in its complementarity with ANT.

Actor

An actor has the ability to make a difference and cannot exist without a network or community. The inseparability between actor and networks solidifies translation and negotiation in the activities of IS/IT solutions (Callon, 1984; Cordella, 2010). ANT emphasises the relationships that are formed between actors during the selection, development, implementation, management, or use of IS/IT solutions in an organisation. Subsequently, these relationships result in networks creation. People, processes, and technologies are actors that integrate to create a network. First, AT takes cognisance of the roles and responsibilities of actors (Karanasios & Allen, 2013), which are thereafter endorsed by relationship and interaction within legitimacy (rules) in a social system, referred to as community (Heeks & Stanforth, 2007). Also, interaction helps to create relationships that influence and have an impact on actors' performances as they carry out activities and events of IS/IT solutions.

Through the relationships, actors build trust among themselves that determines the levels of strength a network possesses. This can also lead to a stable environment. People, process, and technology are the building blocks of IS/IT solutions primarily because the activities and events that are involved in the development and implementation of IS/IT depends on them. The three components basically form networks or communities of

IS/IT activities, from AT and ANT perspectives. I do acknowledge that in ANT, human and non-human elements, such as technology and process, are considered actors and are treated equally. In slight contrast, only people are referred to as actors in AT, while process and technology are classified as tools. In Chapter 10, it is clarified that in AT, human actors, which form part of the social system, make use of tools to create things and events. Irrespective of this, people, process and technology, interactions, and relationships are imperative for an activity. In carrying out the activities of IS/IT solutions, the inseparable nature of the actors leads to conscious and sometimes the unconscious formation of network.

Network

Networks are formed based on events or activities of actors and for actors, in a reproductive process (Law, 1992). This enables negotiation and translation of activities to happen towards a common goal of aligned interests. Thus, ANT is applied as a lens to explore the translation aspect of the theory and to examine and understand the development or implementation of an IT solution within a specific context or environment. From an ANT perspective, the formation and categorisation of networks are understood, as they draw from the traces of the mind through interpretation, defined and shaped by negotiation, interaction, or relationship among actors.

As actors negotiate with one another, interact, and build relationships, different types of networks are created for different activities and events, making the networks heterogeneous. Actors can also be heterogeneous, meaning that they can belong to more than one network. It is the commonality among the actors and the consistency of heterogeneity that make the network or networks stronger and stable. Only then can actors' participation and contribution to a network be influenced by their relationship and interactions with actors in other networks. For example, an actor brings implicit and explicit rules from another network (e.g., software development team) in order to govern and subsequently stabilise its current network (such as a team where the software is being implemented), to reproduce success. The connection between networks promotes a healthy environment and fosters a smooth translation of events and activities concerning allied interest.

Translation

Translation is a process which occurs sequentially from one stage to another until a meaningful resolution is achieved by the actors. Through translation, negotiation takes place and activities are allocated, convened, and executed within networks. This is founded on the interpretation that actors associate with the activities and events under translation. In addition, through the moments of translation, ANT describes how an activity is transformed from one stage to another, which means that an activity exists for translation to take place. According to Andrade and Urquhart (2010), ANT analytic dimensions afford a great deal of insight into the translation of the underlying anatomy of IS activities. In IS studies, translation of activities occurs within a network as actors' relationships and interactions are drawn upon.

Conclusion

This chapter provides a methodological approach on how to complementarily apply AT and ANT as lenses to guide the analysis or interpretation of qualitative data in IS research. Thus, the approach provides and promotes an in-depth interrogative process and facilitates a holistic coverage in the exploration and examination of qualitative data in IS research. The approach helps to address one theory's weaknesses with another's strengths, thereby helping to fulfil the aim and objectives of the study in which the theories are applied, as proposed in this chapter.

The other contributions of this chapter are in its complementary approach and the order-of-use, which are two of the main confusions and challenges that many aspiring researchers encounter in their attempt to combine both AT and ANT in IS research. The distinction is well presented within the premise that AT initiates and focuses on activity, whilst ANT focuses on actors and networks and their heterogeneity, as defined by their aligned interests. Thus, an activity needs to exist before it can be problematised within a network by a focal actor. The order-of-use can also be used to guide deeper interrogation towards eliciting rich data as they manifest, and for critical analysis that would have been impossible if the order were misplaced. The contributions of the chapter are therefore from two main perspectives: theoretical and methodological. Through its addition to existing literature, it advances the use of both AT and ANT to underpin IS research. Methodologically, the chapter's main contribution is from both the model and the framework. This provides an approach on how to combine the two theories in an IS study.

References

An, H. (2010, March). Research on information systems problems based on an ecological angle. In *Challenges in Environmental Science and Computer Engineering (CESCE), 2010 International Conference on* (Vol. 1, pp. 281–284). IEEE.

Andrade, A. D., & Urquhart, C. (2010). The affordances of actor network theory in ICT for development research. *Information Technology & People*, *23*(4), 352–374.

Atkinson, C., & Brooks, L. (2003). StructurANTion: A theoretical framework for integrating human and IS research and development. *AMCIS 2003 Proceedings*, 378.

Baiocchi, G., Graizbord, D., & Rodríguez-Muñiz, M. (2013). Actor-Network Theory and the ethnographic imagination: An exercise in translation. *Qualitative Sociology*, *36*(4), 323–341.

Callon, M. (1981). For a sociology of technological controversies. *Fundamenta Scientiae*, *2*(3/4), 381–399.

Callon, M. (1984). Some elements of a sociology of translation: Domestication of the scallops and the fishermen of St Brieuc Bay. *The Sociological Review, 32*(S1), 196–233.

Callon, M. (1986). Some elements of a sociology of translation: Domestication of the scallops and the fishermen of St Brieuc Bay. In: J. Law (ed.), *Power, action & belief: A new sociology of knowledge?* pp. 196–229. London: Routledge & Kegan Paul.

Cordella, A. (2010). Information infrastructure: An actor-network perspective. *International Journal of Actor Network Theory and Technological Innovation (IJANTTI), 2*(1), 53–72. IGI Global.

Creswell, K., Worth, A., & Sheikh, A. (2011). Implementing and adopting electronic health record systems: How actor-network theory can support evaluation. *Clinical Governance: An International Journal, 16*(4), 320–336.

Dennehy, D., & Conboy, K. (2017). Going with the flow: An activity theory analysis of flow techniques in software development. *Journal of Systems and Software, 133*, 160–173.

Engeström, Y., Miettinen, R., & Punamäki, R. L. (1999). *Perspectives on activity theory*. Cambridge: Cambridge University Press.

Er, M., Kay, R., & Lawrence, E. (2010, April). Information systems and activity theory: A case study of doctors and mobile knowledge work. In *Information Technology: New Generations (ITNG), 2010 Seventh International Conference on* (pp. 603–607). IEEE.

Fenwick, T., & Edwards, R. (2010). *Actor-network theory in education*. London: Routledge.

Gao, P. (2005). Using actor-network theory to analyse strategy formulation. *Information Systems Journal, 15*(3), 255–275.

Hasan, H., Smith, S., & Finnegan, P. (2017). An activity theoretic analysis of the mediating role of information systems in tackling climate change adaptation. *Information Systems Journal, 27*(3), 271–308.

Heeks, R., & Stanforth, C. (2007). Understanding e-Government project trajectories from an actor-network perspective. *European Journal of Information Systems, 16*(2), 165–177.

Hinkelmann, K., & Pasquini, A. (2014, August). Supporting business and IT alignment by modelling business and IT strategy and its relations to enterprise architecture. In *Enterprise Systems Conference (ES), 2014* (pp. 149–154). IEEE.

Iyamu, T. (2013). Underpinning theories: Order-of-use in information systems research. *Journal of Systems and Information Technology, 15*(3), 224–238.

Iyamu, T., & Roode, D. (2010). Theoretical analysis of strategic implementation of enterprise architecture. *International Journal of Actor-Network Theory and Technological Innovation, 2*(3), 17–32.

Jonassen, D., & Land, S. (Eds.). (2012). *Theoretical foundations of learning environments*. London: Routledge.

Kaptelinin, V., & Nardi, B. A. (2006). *Acting with technology: Activity theory and interaction design*. Massachusetts: MIT Press.

Karanasios, S., & Allen, D. (2013). ICT for development in the context of the closure of Chernobyl nuclear power plant: An activity theory perspective, *Information Systems Journal, 23*(4), 287–306.

Latour, B. (1987). *Science in action: How to follow scientists and engineers through society*. Massachusetts: Harvard University Press.

Law, J. (1992). Notes on the theory of the actor-network: Ordering, strategy, and heterogeneity. *Systemic Practice and Action Research, 5*(4), 379–393.

Macome, E. (2008). On implementation of an information system in the Mozambican context: The EDM case viewed through ANT lenses. *Information Technology for Development*, *14*(2), 154–170.

Métrailler, A., & Estier, T. (2014). EVOLIS framework: A method to study information systems evolution records. In *System Sciences (HICSS), 47th Hawaii International Conference on* (pp. 3798–3807). IEEE.

Moawad, N., Liu, K., & El-Helly, M. (2013). Integrating activity theory and semiotics as knowledge elicitation technique. In *Software Engineering, Artificial Intelligence, Networking and Parallel/Distributed Computing (SNPD), 14th ACIS International Conference on* (pp. 137–142). IEEE.

Müller, M., & Schurr, C. (2016). Assemblage thinking and actor-network theory: Conjunctions, disjunctions, cross-fertilisations. *Transactions of the Institute of British Geographers*, *41*(3), 217–229.

Mursu, Á., Luukkonen, I., Toivanen, M., & Korpela, M. (2007). Activity theory in information systems research and practice: Theoretical underpinnings for an information systems development model. *Information Research: An International Electronic Journal*, *12*(3), 1–29.

Peppard, J., & Ward, J. (2004). Beyond strategic information systems: Towards an IS capability. *The Journal of Strategic Information Systems*, *13*(2), 167–194.

Prado, A. B., & Baranauskas, M. C. C. (2012, November). Representing scientific associations through the lens of Actor-Network Theory. In *Computational Aspects of Social Networks (CASoN), 2012 Fourth International Conference on* (pp. 87–92). IEEE.

Rose, J., & Scheepers, R. (2001). Structuration theory and information system development-frameworks for practice. *In the Proceedings of the 9th European Conference on Information Systems (ECIS)* (pp. 217–231).

Sannino, A., Daniels, H., & Gutiérrez, K. D. (2009). *Learning and expanding with activity theory*. Cambridge: Cambridge University Press.

Shaanika, I., & Iyamu, T. (2015). Deployment of enterprise architecture in the Namibian government: The use of activity theory to examine the influencing factors. *The Electronic Journal of Information Systems in Developing Countries*, *71*(6), 1–21.

Shim, Y., & Shin, D. H. (2016). Analyzing China's fintech industry from the perspective of actor–network theory. *Telecommunications Policy*, *40*(2), 168–181.

Tobin, K., Milne, C., & Plakitsi, K. (2013). *Activity theory in formal and informal science education*. Rotterdam, NLD: SensePublishers.

Varga, M. (2003, June). Zachman framework in teaching information systems. In *Information Technology Interfaces, 2003. ITI 2003. Proceedings of the 25th International Conference on* (pp. 161–166). IEEE.

Walsham, G. (1997). Actor-network theory and IS research: Current status and future prospects. In: A. S. Lee, J. Liebenau, & J. I. DeGross (eds.), *Information systems and qualitative research*, pp. 466–480. Boston, MA: Springer.

Walsham, G., & Sahay, S. (1999). GIS for district-level administration in India: Problems and opportunities. *MIS Quarterly*, *23*(1), 39–65.

Wernick, P., Hall, T., & Nehaniv, C. L. (2008). Software evolutionary dynamics modelled as the activity of an actor-network. *IET Software*, *2*(4), 321–336.

Wong, C. M. L. (2016). Assembling interdisciplinary energy research through an actor network theory (ANT) frame. *Energy Research & Social Science*, *12*, 106–110.

Yujie, N., & Xindi, W. (2010, July). Research on the matching of IT strategic planning and business strategy. In *Computer Science and Information Technology (ICCSIT), 2010 3rd IEEE International Conference on* (Vol. 6, pp. 177–181). IEEE.

12 Structuration theory and actor-network theory

Table of contents

Introduction

One of the most important factors in the computing (information systems and technologies (IS/IT)) environment is an understanding of how technology is driven and 'determined' by other components, such as rules, regulations, and resources, within an organisation. Iyamu and Obe (2010) define a computing environment as a structured environment which encompasses software and hardware, including networking, application, and security infrastructure, and provides computer services, support, and maintenance. According to Harrison et al. (2007: 2242), 'technology has the potential to change social and organizational structures; but, at the same time, technology is also affected by these structures in its design, implementation, and use'. The components do not necessarily direct or control the actions of individuals and groups through their existence in one or other materialistic form. Rather, as individuals and groups, they interpret the components enabling and constraining their activities. This happens as they enact the rules and regulations during the allocation of resources, as well as when carrying out the processes and activities.

IS/IT consist of various solutions, such as software, hardware, and network protocols. The solutions are either developed (e.g., software) in-house or purchased from the shelf. Different IS/IT solutions are deployed in organisations primarily to enhance efficiency and effectiveness, and, ultimately, to improve sustainability and competitive advantage. The solutions

DOI: 10.4324/9781003184119-12

are implemented, managed, and used by people through or based on defined rules and processes. This makes the roles of humans and non-humans critical in the use of IS/IT solutions in achieving organisational goals and objectives. Therefore, the components (technology, people, and process) are key and inseparable in the selection, development, implementation, and practice of IS/IT solutions in organisations. For example, technology, people, and process enable and at the same time constrain the deployment and use of IS/IT solutions. The challenges are of such technical and non-technical nature that they are sometimes difficult to identify and understand. Hence, theoretical lenses are required to understand why certain things happen the way they do.

As mentioned in previous chapters, in the last two decades there has been a significant increase in applying sociotechnical theories, such as structuration theory (ST) and actor-network theory (ANT), in information systems (IS) research. According to Johnston (2001), theories provide new ways of theorising sociotechnical systems or information technology (IT) solutions beyond simplifying the tasks, events, and activities that are involved. The use of theories to underpin IS research is primarily intended to investigate, assess, or evaluate the complexities of IT solutions; factors of influence in both development and implementation; and the implications of human interplay, including their intended and unintended consequences in an environment. Kouroubali (2002) affirms that the interplay between 'action' and 'structure' has shaped the development of organisational analysis, and many studies of innovation have focused on characteristics of individuals and organisations as deterministic factors to successful implementation of IT solutions.

Like Chapter 11, this chapter proposes complementarity of two theories, ST and ANT. A guide on how to apply each of the theories is proposed in Chapters 5 and 8, respectively. Attempts to apply ST and ANT complementarily indicate that each of the theories cannot always completely cover the need and objectives of the study in which they are separately employed. Hence, the complementary use of the theories has increased in importance, particularly in empirical studies. From different perspectives, two of the sociotechnical theories, ST and ANT, have been argued to be suitable and have substantial grounds for combination (Greenhalgh & Stones, 2010; Steen, Coopmans & Whyte, 2006). Their efficacy is reasoned to be in their practicality. Thus, interest in complementarily applying the two theories began to brew.

As the idea gained momentum, Atkinson and Brooks (2003) conjoined the two theories to develop a theoretical framework referred to as StructurANTion. This is mainly because the two progenitor theories provide an ontological and epistemological basis for research towards a direction of close proximity. The complementary application of structuration and ANT expands coverage and increases the leverage of understanding how sociotechnical networks, societies of humans and non-humans come into being, persist, and change (Brooks & Atkinson, 2004). Although StructurANTion seems complete and comprehensive, it does not explain its applicability. This angle is of more interest to postgraduates and aspiring researchers. This means that the data collection and subsequent analysis or interpretation of the data is suitably and substantially guided by a combination of the two theories, ST and ANT. The chapter focuses on the complementary application and order-of-use in employing ST and ANT in a study.

Structuration is a form of continual transformation of structures. By implication, it is a production and reproduction of systems. According to Iyamu and Roode (2010), structure in structuration is in totality different from organisational structure. In Giddens (1984), structure is defined as rules and resources. The other theory, ANT, provides a different perspective on the importance of relationships between actors that are both human and non-human. By their very presence, actors work to establish, negotiate, maintain, and revise the construction of organisational networks of aligned interests (Tatnall, & Gilding, 1999). Along the same line or argument, Warzynski (2006: 15) states that 'the theory of translation can be used to reframe interests, negotiate differences, and build an integrated network for leading technological change'.

As established in Chapters 4–9, applying sociotechnical theories in IS research significantly contributes to shaping the findings and the entire study. Complementarily applying ST and ANT in the same study is gradually growing in momentum. This makes the order-of-use vital. Some studies apply ST before ANT, and vice versa. Until now, there seemed to be no model or framework that defined the order-of-use in applying the theories. It could be argued that it depends on the nature and objectives of the study. Despite the objectives of the research, it is significantly essential to circumscribe the order-of-use.

ST is particularly useful in acknowledging and analysing the dialectic between organisational (including the computing environment) and technological structures, as well as human action. ANT is concerned with the analysis of alliances of networks, the varied interests of actors, and the interactions that take place between the actors in the computing environment. An actor-network consists of actors linked together through various interests. ANT emphasises the heterogeneous nature of actor-networks, which consist of and link together both human and non-human elements (Callon, 1991). A core assumption of ANT is that no actor is different in kind from another. Instead, how size, power, or organisation are generated should be studied in an unprejudiced manner (Law, 1992).

As covered in Chapter 8, ST offers a subtle and detailed view of the constitution of social life. Its analytical dimensions through the duality of structure and its associated modalities help in empirical analysis in IS research. However, structuration and other theories do not address the problem of understanding how some social practices become institutionalised (successful), which actor-network does. In addition, ANT helps to follow the actors and understand the study of how networks which include technology are formed and become stable. Thus, the complementary use of both ST and ANT can be substantially benefiting in IS studies.

The remainder of this chapter is structured into five sections. The first section provides the background and motivation for the chapter. This is followed by an overview of the sociotechnical theories, ST and ANT, in the context of IS research. A framework on how to complementarily apply ST and ANT in IS research is proposed and presented in the third section. In the section that follows, the implication of practice is discussed. Finally, the chapter is concluded with an emphasis on its contribution.

Background and motivation for the chapter

Technology such as software, hardware, computer network, and integration provides options which are employed by people to support and enable aims and objectives through processes and procedures. People determine how technologies are used, who uses them, and where and when they are used. The decisions are influenced by factors such as organisational needs, technology-fit, structure, technical skill, as well as the interpretation of the technology. To make meaning and put the influencing factors in perspective, it is critical to examine and understand the actions and interactions that take place during IS and technology deployment and use. This can be achieved by applying one or a combination of theories. The main focus of this chapter is on the complementary application and order-of-use of ST and ANT.

Structuration and ANT concepts have separately been applied in IS research for years and will continue to evolve. Also, few studies have combined the theories in a variety of ways in different IS research. According to Iyamu and Roode (2010), there is no conflict or contradiction in using structuration and ANT in the same research. The authors further argued that the two theories can be applied without comparing and contrasting them. Some of the studies where both theories have been combined include Johnston (2001), which sought to examine and understand how the notion of situated action provides an integration perspective for understanding the theories of structuration and ANT; Atkinson and Brooks (2003) interpreted ST and ANT into a 'new' theory which they termed StructurANTion. Brooks and Atkinson (2003) employed StructurANTion to analyse processes and activities in an IT solution, a case of decision-making, and Iyamu and Roode (2010) applied both theories separately and complementarily to investigate and understand how IT strategy can be developed and implemented in two different organisations.

Structuration and ANT theories are complementary towards phenomena in the social construction in organisations. Brooks and Atkinson (2004) provide comprehensive rationales for combining ST and ANT, based on which the authors developed a framework termed StructurANTion. It demonstrates the growing interest in furthering rigour in empirical research in IS. However, Jones and Karsten (2008) view the integration of ST and ANT to incorporate more distinctive components. What is missing in Brooks and Atkinson and other studies mentioned above are two momentous factors: (1) explanation and decomposition in the form of a guide on how to complementarily apply the theories; and (2) the order-of-use in applying the theories. These are the primary concern and focus of this chapter. It aims to contribute to the development of IS research and body of knowledge in general, from these angles.

In a study where theories are combined, the order-of-use is key and highly significant, as it has an influence on the findings, from data collection to analysis and interpretation of empirical studies. It raises some fundamental questions such as which theory should be applied before the other and what influences the order-of-use? Also, can people and activities form networks, or can people and activities operate within networks? The factors are inseparable. Hence, it is important to define and formalise the order-of-use of the underpinning theories. This chapter addresses these questions.

Applying structuration and ANT to underpin research helps to foster and gain an understanding of the interaction that happens during technology deployment. It involves prefixing the inquiry with what, when, why, where, and how technologies are developed, implemented, and used. According to Orlikowski and Robey (1991), ST is applied to help understand the relationship between IT solutions and organisations. Walsham (1997) argued that ANT can be useful in a wide range of IS research that offers considerable potential for increased understanding of the sociotechnical systems. ANT concentrates on how things get done, to the detriment of how broader social structures shape socio-material practices. It gives interesting accounts of local contingencies and material arrangements (Latour, 1991).

ST, ANT, and IS

The theories ST and ANT are mostly employed from the perspectives of duality of structure and moments of translation, respectively, in IS research. Both theories are usually applied within the frame of interpretivist and qualitative studies. The interpretive study is characterised by the need to gain a better understanding of the world as it is, and to have deeper insight of the fundamental nature of the real world at the level of subjective experience. It seeks explanation and clarification within the frame of reference from participants' action and experience, rather than the observer's view. Thus, the main aim of interpretivist research is often to understand, rather than to predict. Therefore, it allows the researchers to interact with humans, thereby changing the perceptions of both parties within context, and on the phenomenon being studied. Both ST and ANT allow for social interaction as a means of assessing, explaining, and understanding phenomena.

Structuration offers a subtle and detailed view of the constitution of social life. Pozzebon and Pinsonneault (2001) argued that ST recognises and accommodates both subjective and objective dimensions in the field of IS research. In structuration, shared knowledge creates potentialities that influence actors' actions, which in turn reinforce existing expectations (Winter & Saunders, 2019). Agency and structure are mutually constitutive (Chanias, Myers & Hess, 2019), meaning that actors involved in IT solutions have a choice between recursive and adaptive approaches. From the perspective of structuration, organisational change is the joint efforts of the actions of individuals interacting with structures (rules and resources) within an institution. These structures both enable and constrain the daily actions and thought processes of people, but do not wholly determine them.

From a structuration viewpoint, individual choices are not independent of the structures within which they take place, but they can move towards maintaining, reinforcing, changing, or even destroying them. According to Giddens (1984), one of ST's contributions relates to the reciprocal, mutually reinforcing nature of agents and structure interaction. Conceptualising a structure as flexible and recognising that rules and resource control mechanisms are continually reinforced and/or redefined through agent action enable a dynamic view of ethics within the organisation. One of the main propositions of ST is the duality of structure, which implies that the rules and resources drawn upon in the production and reproduction of social action are at the same time the means of system reproduction (Giddens, 1984).

ANT does not distinguish between human and non-human agents and it rejects distinctions between the human and non-human factors that form the basis of most IS studies. The ANT perspective attempts to explain and interpret social and technological evolution, using neither technical-material nor social reductionism (Meneklis & Douligeris, 2007). Rather, it incorporates principles that integrate both humans and non-humans into the same conceptual framework. Also, ANT systematically avoids methodological dualism. This is the drawing of a priori distinction between technical and non-technical elements (Bloomfield & Vurdubakis, 1997). This simplifies the complexity of assumption and pretence of dealing with two separate, but related, ontological domains. They should rather be regarded as phases of the same essential action (Latour, 1991).

The actor network's focus on the heterogeneous networks of aligned interests is ostensibly to endorse negotiation and transformation, shifting from one stage to another. The treatment of both human and non-human actors by ANT is based upon three principles: agnosticism, generalised symmetry, and free association (Callon, 1986): agnosticism (meaning analytical impartiality) is demanded towards all the actors involved in the study, whether they are human or non-human; generalised symmetry offers to explain the conflicting viewpoints of different actors in the same terms by use of an abstract and neutral vocabulary that works the same way for human and non-human actors; and free association requires the elimination and abandonment of all a priori distinctions between the technological or natural and the social aspect.

Duality of structure

The duality of structure focuses on the processes through which legitimating structures evolve and are reconstituted by action. In structuration, structure is constructed by interacting with individuals whose activities are enabled and constrained by structure even as they form the patterns that are then recognised as structure (Hatch, 1997). The interplay between individuals and structures is often conceived as the duality of structure and is one of the sources of reasoning leading to the acceptance that the organisational changes that emerge in similar situations are not completely predictable.

Structuration's analytical dimension is through the duality of structure and its associated modalities could be very detailed and helpful for analysis or interpretation in empirical studies in the field of IS. This is predominantly because human actions are composed of structures of meaning, power, and moral frameworks. Giddens (1984) offers three modalities that link human action to social structure. These are interpretative schemes (the structure of signification), resources (the structure of domination), and norms (the structure of legitimation). Interpretative schemes are a form of standardisation and shared stocks of knowledge that people use to interpret behaviour and events, and thereby achieve meaningful interaction among individuals and groups within context; resources are either authoritative or allocative. They are the means for exercising power, accomplishing goals, and realising intentions in situations, and norms are rules for sanctioned or appropriate conduct, defining the legitimacy of interaction within a setting's moral order.

Through agents' interaction, social structures are created which, in turn, influence the actions and interactions of humans. This is an element of duality, which signifies that human actions are not static but have a transformative capacity to produce and reproduce structures (Vyas, Chisalita & Dix, 2017). Transformative capacity is an enactment of power and ability to intentionally carry out actions. It is in this vein that capability or power is knowingly or unknowingly employed, a characteristic that is distinct about human beings. Hence, agents are said to have acted with agency if they have consciously carried out an activity. Giddens (1984) emphasises that consciousness is the ability of agents to execute their actions in an event.

As agents communicate, they draw on interpretive schemes for sense making to assess signification, and the interaction is used to reproduce and modify structure in a social system. In the process, as agents interact and communicate, they tap on their power. This determines the extent to which actions of agents can influence structures. In structuration, power is viewed as the cogent ability of an agent to make a difference and in the process reproduce actions. Agents exercise their power over others through domination by using resources or facilities that are legitimately sanctioned within the social system.

Structure is therefore process- and activity-dependent. It is both the medium and the outcome of a process of 'Structuration' – the production and reproduction of practices across time and space. This process is what Giddens called the 'double hermeneutic', the double involvement of individuals and institutions. Put perhaps more clearly, we create society at the same time as we are created by it (Giddens, 1984: 14). This results from modalities that link particular types of interaction with particular structural elements (Giddens, 1984). The duality of structure helps in understanding the contextual dynamics within which ethical dilemmas are resolved. Structures embody the ethical norms which influence actions. Actions by agents may lead to changes in how rules and resources influence interactions and reinforce the norms upon which these interactions are based. Once in place, the new process fosters further changes through interacting modalities.

As stated in Chapter 8, structuration is increasingly applied in IS studies owing to its relative closeness to IS phenomena. In many quarters, the theory is deemed useful and appropriate to underpin IS research in providing informed accounts of social practices in IS and technology, a field dominated by technical and non-technical considerations. Its sociological theorising and analysis, including interpretation, offer rich insights about social action that informs practices of IT solutions.

Moments of translation

An actor network consists of actors linked together through various interests. ANT gives privilege to neither human nor non-human factors. Instead, it incorporates a 'principle of generalised symmetry', where both human and non-human elements are considered actors or actants and both elements are capable of affecting each other. ANT emphasises the heterogeneous nature of actor networks which consist of and link together both human and non-human elements (Callon, 1991). ANT is also concerned about how people and objects are brought together in stable, heterogeneous networks of aligned interests through processes of translations and negotiations (Callon & Law, 1989). Latour (1996) emphasises that heterogeneous networks overcome issues related to identity and avoid arbitrary dichotomies and structures. Heterogeneity reflects in different organisational principles and objectives that are in simultaneous action.

The theory attempts to answer how a diverse group of actors reaches agreement; that is, how a social order establishes a certain degree of stability or exhibits structural properties. In ANT, stability is the result of the social process aligning an initially diverse collection of interests into an acceptable stability which has elements of irreversibility (Callon, 1991). Latour (1996) suggests that each element in a network is simply defined by the heterogeneous list of its associates. Entities are hence defined through their existence and their specific associations with other entities, and they are not substitutable. The elements, whether human or non-human, within a network acquire power through the number, extensiveness, or stability of the connections routed through them. Although there has been a debate about the philosophical validity of giving equal status to human beings and non-human artefacts, the notion is perhaps less revolutionary in the field of IS (Walsham, 1997).

Translation implies transformation, which refers to how actors engage with other actors through negotiation to generate ordering effects (Callon, 1986; Law, 1992). Callon (1991) emphasises that translation goes beyond the traditional definition of action as it deals with mutual understanding and inscription. Thus, actors negotiate or manoeuvre others' interest to their own with the aim of enrolling actors into the network. Such inscriptions prescribe a programme of action for other actors, although they can vary in strength and flexibility (Hanseth, Monteiro & Hatling, 1996). Problematisation, interessement, enrolment, and mobilisation are the four moments of translation.

The first stage of the translation process, problematisation, describes a system of alliances, or associations between entities, thereby defining the identity and the intention. The idea of problematisation is to foster relationships and to allocate and reallocate tasks which give power to involved actors. The second stage, interessement, describes how the allies are locked into negotiation of predictive outcome as defined by the focal actor. This is followed by enrolment, a stage that distinguishes itself from interessement in that not all the interested parties or actors participate in the activities of the network. Thus, the enrolment is the moment that helps to consolidate mutual consent as defined by the focal actor within legitimated rules and regulations. The final stage is mobilisation, a stage where there are self-appointed spokespersons intended to help strengthen and stabilise the network.

The relational and process-oriented sociology of ANT provides a solid theoretical underpinning for unravelling and understanding the impact human and non-human elements have on each other (Tatnall & Gilding, 1999). ANT offers a precise and non-functionalist account of how actors become established as powerful through the stability of the networks that pass through them. The actor (human or non-human) that is an obligatory passing point in a network has 'power', and the more networks the actor belongs to, the more 'power' that actor has. As a result, over time, the ability of an actor to act effectively on a larger scale becomes established (Callon, 1991).

Like ST, applying ANT in IS research has been growing in popularity over the years, as denoted in Table 10.1. Thus, there has been increasing interest in applying ST and ANT as lenses in IS research in a variety of ways. Based on the objectives of the chapter, a framework is developed. It is intended to help with the complementary application of the theories of ST and ANT in IS studies.

Framework for complementary use of ST and ANT

The computing environment, which constitutes IS and technology, consists of both technical and social factors, which are primarily technology, people, and process – see Figure 12.1. These components are inseparable, and they depend on as well as complement each other in the activities of the computing environment. As mentioned earlier in the chapter, people determine how the technologies are used, who uses them, and where and when the technologies are used to enable and support activities and events in an organisation. These decisions are influenced by the organisational structures and needs, including the technical skill and interpretation of the processes that are involved in a technology (IT solution).

The selection, development, implementation, or practice of IT solutions is conducted through processes formulated by people. As such, the success or failure of technology deployment or use greatly depends on the people involved and who undertake the responsibilities of deploying the IT solution and enforcing the processes of doing so. Orlikowski and Gash (1994) argue that an understanding of people's interpretations of a technology is critical to understanding their interaction with it. To interact with technology, people must have knowledge of it to make sense of it, and in the process of making sense of it, develop particular interests, assumptions, expectations, and purpose of the technology, which then serve to shape subsequent actions towards it. Hence, it is essential to identify with why and how people act on the basis of their interpretations and understanding of an IT solution. This drives and determines the capabilities of the technology, the process of using the technology, and the outcomes of the technology or what it produces.

In the context of the chapter, process includes procedures, rules, and regulations. Within process, people execute technology for a variety of objectives. The process is thereby produced and reproduced in attempts to achieve objectives. The elements of process form networks within which

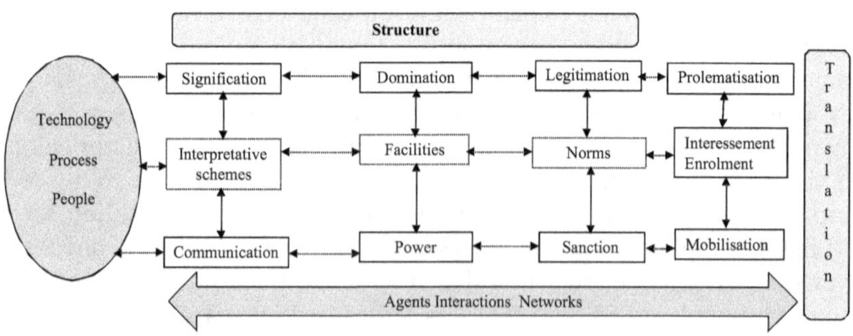

Figure 12.1 Interaction between sociotechnical agents and actors.
Source: Author.

interaction takes place. Process is the component of technology that opens up the possibilities for the development, implementation, and use of technology. According to Mooney, Gurbaxani, and Kraemer (1996), process is a specific ordering of work activities across time and place, with a beginning, an end, and clearly identified inputs and expected outputs.

As depicted in Figure 12.1, the three components, technology, people, and process, as explained above, deny the separability of the social and the technical elements in a computing environment, which consists of a network of actors and structure (rules and resources). The same goes for IS/IT solutions that are continually assessed and evaluated through IS research.

Structuration does not allow the relationships between people and technology to be probed. Consequently, this makes it difficult to employ the theory in the analysis of interaction between individuals, groups, and technology. ANT bridges this gap through its examination of the relationships between human and non-human actors. It focuses on the sociotechnical networks that humans create to accomplish individual or group goals and interests. ANT emphasises that no one acts alone, as there are important roles played by other actors, which are of technical and non-technical nature. Both the development and implementation of an IT solution require the process of negotiation and shifting of translation and interest.

Groups and individuals are interested in subjectivism or objectivism within the frame of consciousness, unconsciousness, or practical consciousness. Signification is the value associated with entities that are driven by moments of translation (problematisation, interessement, enrolment, and mobilisation). Due to the inseparable nature of the components, technology, people, and process, the interaction that happens between them is coerced. Figure 12.2 depicts the connotation of the interaction that takes place between structure and actor from the perspectives of ST and ANT. The figure is further discussed below. The interaction happens within a network, following a process and using available resources.

As shown in Figure 12.2, there is an anecdotal relationship between structure and actor. From one angle, the relationship entails the involvement of an agent who, through innovation, creates a process that manifests to the power an actor has. From another perspective, structure guides the creation of process, which gives power to an agent to innovate. Consequently, structure and actor create or define power relationships between agents and their involvement with innovation and process. Thus, the actor provides a platform which allows for the analysis that can draw on the relationship of both humans and non-humans within a network or heterogeneous networks. This makes it critical to assess problematic sets of explicit and implicit assumptions made from the human perspective on non-humans' existence. ANT is predominantly applied as a lens for analysing or interpreting data in order to gain an understanding of the spheres of influence and the power of actors.

Structure and agents endorse power to determine, execute, and manifest through processes to enable and support innovation. This is a recursive

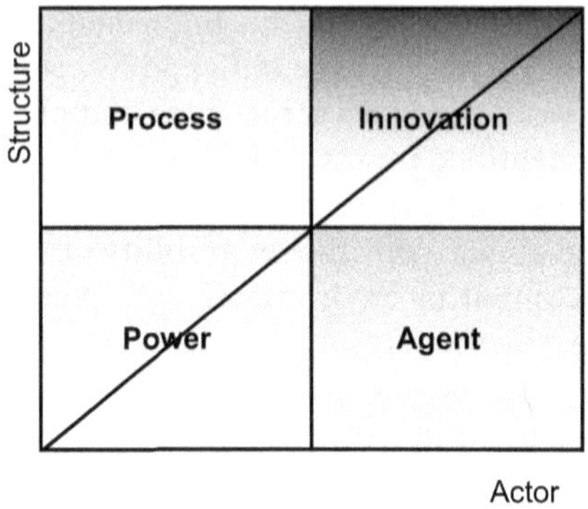

Figure 12.2 Structure and actor.
Source: Author.

approach that connects subjects and social objects, using the experiences of individual actors across space and time. Human social activities, such as self-reproducing items, are recursive in nature. That is to say, they are not brought into being by social actors but are continually recreated by them via the very means whereby they express themselves *as* actors. Structuration thus recasts the classical categories of structure and agency as a duality in a dialectical framework. The contribution of ST is that it provides a means for understanding how social institutions are produced and reproduced over time within context. The theory of structuration suggests that human actions simultaneously condition and are conditioned by organisational properties in social contexts.

Based on the discussion above, including the illustration of Figures 12.1 and 12.2, it is as clear as it was before now that ST and ANT can be complementarily applied. The fresh perspective is that both theories can be applied in the order of micro and macro interconnected levels for interpretive analysis of data in IS studies, as shown in Figure 12.3. This is the part that has posed challenges to many researchers, including postgraduates. Also, this brings about a guiding principle. The different levels ensure consistency in their applicability in IS studies.

The first, the micro level, addresses issues of IS importance to the case (organisation) in a study, as well as of the relationship between human and non-human actors through process and within structure, as illustrated in Figure 12.3. This level provides a better understanding of the social context of the study. The purpose of the first level of analysis is to examine and gain an insight into the social context. Technology and process are built within a social context because society exists. At this level, the analysis considers the environment in which the study is conducted. This is to examine why and how both the internal and the external factors or variables influence

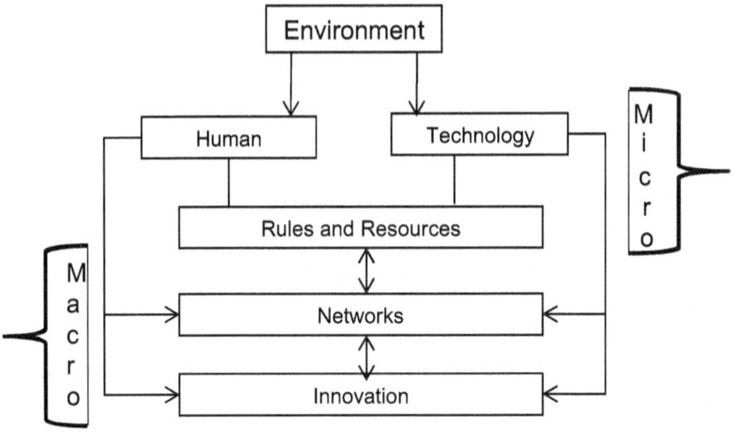

Figure 12.3 Theoretical framework for complementary application of ST and ANT.
Source: Author.

the activities of the IT solution that is being studied. This helps to assess the roles of the social context and the controlling effect technical factors have on the IT solution in the organisation.

The macro, the second level of data analysis or interpretation, is intended to gain a better comprehension of the factors that influence and have the ability to challenge the status quo or inflame a difference at the actors' level and within the context. At this level, knowledge about the impact on activities, processes, and structures is sought and gained. Also, at this level, the environment (case) in which the research is conducted is considered. Internal and external factors, as well as the demands of the wider organisational and human (social) contexts, influence the settings of the case environment.

Figure 12.3 clearly shows that the macro level examines the relationship between humans and technological artefacts or solutions. The intricacies of the choice of (innovative) technology and how they are applied are sought. Understanding is gained of how this relationship is fostered and managed by and with the available resources within the rules of the network. How initiatives are formulated and developed, including how technology is applied, by whom, where, when, and why can also be examined. Thus, the macro level of analysis addresses issues of IS importance to an organisation, as well as the relationship between human and non-human actors, within process and structure. It therefore helps to better explain the social context of the phenomenon being studied.

IS and technologies (IT solutions) are products of human actions for specific goals and objectives. Similarly, the networks are formed or constituted within context and rules, using the available resources. This makes the translation of the solutions to points of acceptability (sanction) whereby they are legitimised by the stakeholders (human actors). Problematisation does not happen in a vacuum but within a network. Also, the networks are constituted within and based on structure (rules and resources). The structure defines the boundaries and the capacity of technical and non-technical entities within which a network exists and, within it, produces and reproduces actions and activities, such as the development and implementation of IT solutions.

It is at the macro level that information about the sequence of innovation can be elicited, making it easier to understand the bytes and bits of elements that are involved in the innovation of an IT solution, in both development and implementation. This means that the macro analysis can help reduce the process and composition of innovation of IT solutions to a grander scale. Also, it embeds a relative deterministic scale of IT solution innovation that draws on distinction between phenomena. The macro level of analysis illuminates whether an innovation was based on the conscious or unconscious action of human actors.

The macro levels help to elucidate the interaction between people and how it is mediated through objects of various kinds, including how such interactions are in turn mediated through additional networks of objects and people. In addition, the macro level of analysis helps reveal how the different networks participate in shaping the social interplay in the delivery of IT solutions. Hence, the view in ANT is that a particular order is an effect generated by heterogeneous means between the micro and macro levels. An actor can therefore be produced from or as an effect of these heterogeneous relations between people and objects, and an actor is also, always, part of a network (Law, 1992).

The translation of interests is roughly analogous to persuasion and the negotiation of common interests by human actors. Between humans and objects, translation occurs during design when the object is imbued with its purpose, programme, or script, as to how it would interact with or affect other actors (Akrich, 1992). Further translation takes place between the object and the actors it encounters as the initial programme or script is altered through interaction.

Implications for practice

The use of theories to underpin IS studies has implications for practice in that the study may attain more rigour than relevance. Thus, as shown in Figure 12.4, it is necessary to find a balance between rigour and relevance. Rigour and relevance have been hugely debated in the last two decades. A few of the voices in the debate include Benbasat and Zmud (1999), Gray (2001), and Lee (1999). This chapter has the intention neither to get into the debate nor to trivialise it. This could be an area for further work in the order-of-use of theories in IS research.

Many qualitative studies in IS are questioned about the depth of rigour. This is as a result of little or lack of intensity, synthesis, and appropriate

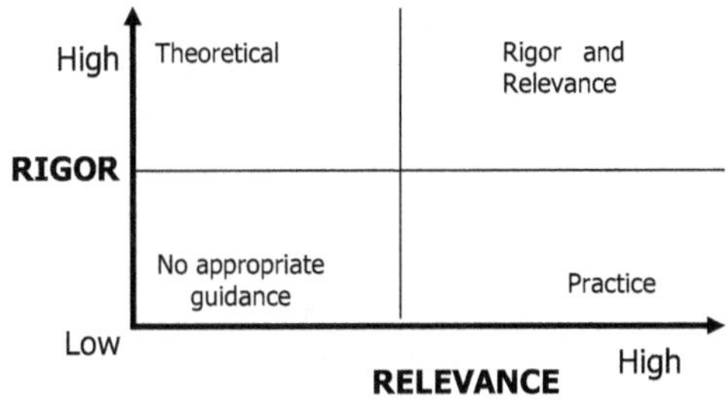

Figure 12.4 Implication for practice.
Source: Author.

guidance in theoretical assumptions. On the other hand, some research articles which consist of practical depth and the application of a real-world situation but lack rigour face criticisms from IS practitioners. Khazanchi and Munkvold (2003) argued that the better the theories about IS phenomena, the greater their chances of having a long-term impact on IS practice. This requires a balance between theoretical assumption and relevance in studies, as illustrated in Figure 12.4.

Though social contexts within phenomena under which a study is conducted require in-depth theoretical assumptions and underpinning, they do also require a good understanding of a real-world practice. This has implication and impact on the order-of-use of theories to underpin a study, as it is characteristically relied upon to help exhume the dependence and relationships which exist between (i) human actors; (ii); non-human actors; and (iii) human and non-human actors within the social context.

Applying theories to underpin IS research is one thing. Selecting the most appropriate theories is another rudimental factor. Intricately, the order-of-use in a complementarity of theories helps to achieve rigour in a study. This is primarily because the theories guide the researcher from data collection to the analysis and interpretation stages. The symmetric sequence which the applicability of complementarity of theories brings into a study makes it holistic and assists in avoiding gaps, from the perspectives of rigour and relevance. Otherwise, the scrupulous and meticulous nature of rigour will continue to dominate claims for practice, as shown in Figure 12.4. The use of theories to underpin studies is critical, as there is always the need to find a balance between rigour and relevance. In IS research, rigour and relevance are vital to building consistency and cohesion as well as creditability and solidity into a study.

Conclusion

This chapter presents a rationale and proposes how structuration and ANT can be complementarily applied in an IS research. It also demonstrates order-of-use, thereby reducing complexity in the course of complementarity.

It therefore helps to understand how theories can be adopted in the order of methodological value. The way data is collected, organised, analysed, and interpreted is influenced and shaped by the order-of-use in applying theories. The importance of the order-of-use is to create consistency, predictability, and uniformity in analysis. This has an impact on the findings. The chapter constitutes a learning curve and is expected to benefit the computing industry and IS researchers on the importance of complementarily applying ST and ANT to underpin studies.

No organisation has total power to determine what the choice(s) of an actor will be in a particular circumstance. On the one hand, actor and structure, by virtue of their interaction, recursively produce and reproduce IT solutions in organisations. On the other hand, actor and structure enable and at the same time are constraints. Complementarity of structuration and ANT as lenses traces the trajectories of both enablement and constraining factors in the development, implementation, or practice of IT solutions in an organisation. Thus, the theories can be used as a way of thinking and speaking about the phenomena being studied. It also helps to understand the consequences and implications of applying the lenses in a manner different from this particular order.

References

Akrich, M. (1992). The de-scription of technical objects. In: W. Bijker and J. Law (eds.), *Shaping technology/building society: Studies in sociotechnical change*, pp. 205—224. Cambridge: MIT Press.

Atkinson, C., & Brooks, L. (2003). StructuANTion: A theoretical framework for integrating human and IS research and development. In: J. Ross and Galleta (eds.), *Proceedings of the ninth Americas conference on information systems*, pp. 2895–2902. Tampa. Florida.

Benbasat, I., & Zmud, R. W. (1999). Empirical research in information systems: The practice of relevance. *MIS Quarterly, 23*(1), 3–16.

Bloomfield, B., & Vurdubakis, T. (1997). Visions of organization and organizations of vision. *Accounting, Organizations and Society, 22*(7), 639–668.

Brooks, L., & Atkinson, C. (2003). StructurANTion in research and practice: Representing actor networks, their structurated orders and translations. In: B. Kaplan, D. Truex, D. Wastell, T. Wood-Harper, & J. I. DeGross (eds.), *Information systems research: Relevant theory and informed practice*, pp. 389–409. Boston, MA: Springer.

Callon, M. (1986). Some elements of the sociology of translation: Domestication of the scallops and the fisherman of St Brieuc Bay. In: J. Law (ed.), *A new sociology of knowledge, power, action and belief*, pp.196–233. London: Routledge.

Callon, M. (1991). Techno-economic networks and irreversibility. In: J. Law (ed.), *A sociology of monsters. Essays on power, technology and domination*, pp. 132–164. London: Routledge.

Callon, M., & Law, J. (1989). On the construction of sociotechnical networks: Content and context revisited. *Knowledge and Society: Studies in the Sociology of Science Past and Present, 8*, 57–83.

Chanias, S., Myers, M. D., & Hess, T. (2019). Digital transformation strategy making in pre-digital organizations: The case of a financial services provider. *The Journal of Strategic Information Systems*, *28*(1), 17–33.

Giddens, A. (1984). *The constitution of society: Outline of the theory of structuration.* Cambridge: John Polity Press.

Gray, P. (Ed.) (2001). Special issue on relevance. *Communications of the AIS*, *6*.

Greenhalgh, T., & Stones, R. (2010). Theorising big IT programmes in healthcare: Strong structuration theory meets actor-network theory. *Social Science & Medicine*, *70*(9), 1285–1294.

Hanseth, O., Monteiro, E., & Hatling, M. (1996). Developing information infrastructure: The tension between standardization and flexibility. *Science, Technology, & Human Values*, *21*(4), 407–426.

Harrison, T, Pardo, T, Gil–Garcia, J, Thompson F., & Juraga, D. (2007). Geographic information technologies, structuration theory, and the world trade center crisis. *Journal of the American Society for Information Science and Technology*, *58*(14), 2240–2254.

Hatch, M. J. (1997). *Organization theory: Modern, symbolic, and postmodern perspectives.* Oxford, New York: Oxford University Press.

Iyamu, T., & Obe, O. (2010). Components of the computing environment. *Computer Sciences and Telecommunications Journal*, *26*(3), 142–146.

Iyamu, T., & Roode, D. (2010). The use of Structuration and Actor Network Theory for analysis: A case study of a financial institution in South Africa. *International Journal of Actor-Network Theory and Technological Innovation*, *2*(1), 1–26.

Johnston, R. (2001). Situated Action, Structuration and Actor-Network Theory: An integrative theoretical perspective. *In the Proceedings of the 9th European Conference on Information Systems Bled*, Slovenia, June 27–29 (pp. 232–242).

Jones, M. R., & Karsten, H. (2008). Giddens's structuration theory and information systems research. *MIS Quarterly*, *32*(1), 127–157.

Khazanchi, D., & Munkvold, B. E. (2003, January). On the rhetoric and relevance of IS research paradigms: A conceptual framework and some propositions. In *36th Annual Hawaii International Conference on System Sciences, 2003. Proceedings of the* (pp. 1–10). IEEE.

Kouroubali, A. (2002). Structuration theory and conception-reality gaps: Addressing cause and effect of implementation outcomes in health care information systems. *Proceedings of the 35th Hawaii International Conference on System Sciences.*

Latour, B. (1991). The impact of science studies on political philosophy. *Science, Technology, & Human Values*, *16*(1), 3–19.

Latour, B. (1996). On actor-network theory: A few clarifications. *Soziale Welt*, *47*, 369–381.

Law, J. (1992). Notes on the theory of the actor-network: Ordering, strategy, and heterogeneity. *Systems Practise*, *5*(4), 379–393.

Lee, A. S. (1999). Rigor and relevance in MIS research: Beyond the approach of positivism alone. *MIS Quarterly*, *23*(1), 29–34.

Meneklis, V., & Douligeris, C. (2007). Enhancing the design of e-government: Identifying structures and modelling concepts in contemporary platforms. *ACM*, 208–216.

Mooney, G. H., Gurbaxani, V., & Kraemer, L. K. (1996). A process oriented framework for assessing the business value of information technology. *ACM SIGMIS Database*, *27*(2), 68–81.

Orlikowski, W. J., & Gash, D. C. (1994). Technological frames: Making sense of information technology in organizations. *ACM Transactions on Information Systems (TOIS)*, *12*(2), 174–207.

Orlikowski, W., & Robey, D. (1991). Information technology and the structuring of organizations. *Information Systems Research*, *2*(2), 143–169.

Pozzebon, M., & Pinsonneault, A. (2001). Structuration theory in the IS field: An assessment of research strategies. *In the Proceedings of the 9th European Conference on Information Systems,* Slovenia, June 27–29, pp 205–217.

Steen, J., Coopmans, C., & Whyte, J. (2006). Structure and agency? Actor-network theory and strategic organization. *Strategic Organization*, *4*(3), 303–312.

Tatnall, A., & Gilding, A. (1999). Actor-network theory and information systems research. *In the Proceeding of the 10th Australian Conference on Information Systems, (ACIS),* Wellington, Victoria University of Wellington.

Vyas, D., Chisalita, C., & Dix, A. (2017). Organizational affordances: A structuration theory approach to affordances. *Interacting with Computers*, *29*(2), 117–131.

Walsham, G. (1997). Actor network theory and IS research: Current status and future prospects. In: A. S. Lee, J. Liebenau, & J. I. DeGross (eds.), *Information systems and qualitative research*, pp. 466–480. London: Chapman & Hall.

Warzynski, C. (2006). An actor-network approach to leading technological change: Implementing a new technology at a prominent U. S. Research University. In: A. M. Poorthuis (ed.), *The network approach: Building organisations and society.* Amsterdam: Van Gorcum.

Winter, S. J., & Saunders, C. (2019). The personal in the policy cascade. *Journal of the Association for Information Systems*, *20*(11), 247–265.

Index

Note: **Bold** page numbers refer to tables and *italic* page numbers refer to figures.

For Product Safety Concerns and Information please contact our EU representative GPSR@taylorandfrancis.com
Taylor & Francis Verlag GmbH, Kaufingerstraße 24, 80331 München, Germany

www.ingramcontent.com/pod-product-compliance
Lightning Source LLC
LaVergne TN
LVHW021126300926
849160LV00010B/881
9781032026015